Relational Mechanics

Andre K.T. Assis

Apeiron
Montreal

Published by Apeiron
4405, rue St-Dominique
Montreal, Quebec H2W 2B2 Canada
http://redshift.vif.com

Dr. Andre Koch Torres Assis
Institute of Physics
State University of Campinas
13083-970 Campinas - SP, Brazil
E-mail: assis@ifi.unicamp.br
Homepage: www.ifi.unicamp.br/~assis

First Published 1999

Canadian Cataloguing in Publication Data

Assis, Andre Koch Torres, 1962-
    Relational mechanics

Translation of Mecanica Relacional
Includes bibliographical references and index.
ISBN 0-9683689-2-1

    1. Mechanics. 2. Gravitation. 3. Inertia (Mechanics). I. Title.

QC125.2.A88 1999   531   C99-900382-8

Front cover:
Newton's bucket experiment. The water rises in the second case
because of its rotation. But: rotation relative to what? This is the
central theme of this book.

Back cover:
Galileo's free fall experiment. Why do two bodies of different
weight fall in vacuum with the same acceleration toward the
surface of the earth? Relational Mechanics offers a simple
explanation for this remarkable fact.

In memory of Isaac Newton
who paved the way for past, present and future generations

Contents

Acknowledgments

To the undergraduate and graduate students who followed our courses on Relational Mechanics, Mach's Principle, Weber's Electrodynamics and Cosmology, for the many constructive remarks they presented. To our undergraduate and Ph.D. students who are developing researches in these areas. To our friends who are giving us support to continue this work. In special we wish to thank those to whom we gave a first version of this book: C. Roy Keys, Thomas E. Phipps Jr., A. Ghosh, D. Roscoe, A. Martin, R. de A. Martins, M. A. Bueno, W. M. Vieira, M. A. de Faria Rosa, A. Zylbersztajn, D. S. L. Soares, J. I. Cisneros, H. C. Velho, D. Gardelli, J. A. Hernandez and J. E. Lamesa. The suggestions and ideas which we received contributed greatly to the improvement of the book.

To the following Publishers and individuals for granting permission to quote from their publications: University of California Press (I. Newton, Mathematical Principles of Natural Philosophy, 1934, Cajori edition); Open Court Publishing Company, a division of Carus Publishing Company, Peru, IL (E. Mach, The Science of Mechanics, 1960, translated by T. J. McCormack); Everyman's Library, David Campbell Publications Ltd. (G. Berkeley, Philosophical Works - including De Motu and A Treatise Concerning the Principles of Human Knowledge, 1992); Dover Publications (A. Einstein, H. A. Lorentz, H. Weyl and H. Minkowski, The Principle of Relativity, 1952); Cambridge University Press (J. B. Barbour, Absolute or Relative Motion?, 1989); Albert Einstein Archives, The Hebrew University of Jerusalem, Israel (A. Einstein, The Meaning of Relativity, 1980), Dr. Peter Gray Lucas, in name of Dr. H. G. Alexander (The Leibniz-Clarke Correspondence, edited by H. G. Alexander, 1984); Birkhauser Boston and Dr. J. B. Barbour for his translations of E. Schrödinger, H. Reissner, A. Einstein, B. and I. Friedlaender (Mach's Principle - From Newton's Bucket to Quantum Gravity, edited by J. B. Barbour and H. Pfister, 1995).

To the Center for Electromagnetics Research, Northeastern University (Boston, USA), which received us for one year in which we had the first idea to write this book and discussed its contents with some friends. To the Institutes of Physics and Mathematics and to the Center of Logics, Epistemology and History of Sciences of the State University of Campinas - UNICAMP (Brazil), that gave the necessary support to undertake this work.

Above all, to my parents, my wife and children for helping me grow as a human being.

Preface

This book presents Relational Mechanics, a new mechanics which implements the ideas of Leibniz, Berkeley, Mach and many others. Relational mechanics is based only on relative quantities, such as the distance between material bodies, their relative radial velocity and relative radial acceleration. In this new mechanics the absolute concepts of space, time and motion do not appear. The same can be said of inertia, inertial mass and inertial frames of reference. When we compare relational mechanics with Newtonian mechanics, we will gain a new and clear understanding of these old concepts. Relational mechanics is a quantitative implementation of Mach's ideas utilizing a Weber's force law for gravitation. Many people have contributed to its development, including Erwin Schrödinger.

This is the first time such a book has been written, bringing together all the features and characteristics of this new world view. This allows it to be seen in its proper light, and a comparison with old worldviews is easily accomplished.

Considerable emphasis is placed on Galileo's free fall experiment and on Newton's bucket experiment. These are some of the simplest experiments ever performed in physics. Despite this fact, no other experiment has had such far-reaching consequences for the foundations of classical mechanics. An explanation of these two experiments without utilizing the concepts of absolute space and inertia is one of the major accomplishments of relational mechanics.

In order to show all the power of relational mechanics and put it in perspective, we first present Newtonian mechanics and Einstein's theories of relativity. We address the criticisms of Newton's theory made by Leibniz, Berkeley and Mach. Then we present relational mechanics and show how it solves all these problems quantitatively with a clarity and simplicity unsurpassed by any other model. We also discuss the history of relational mechanics in detail, emphasizing the achievements and limitations of all major works along these lines. In addition, we present several notions which are beyond the scope of Newtonian theory, such as the precession of the perihelion of the planets, the anisotropy of an effective inertial mass, the adequate mechanics for high velocity particles, *etc.* Experimental tests of relational mechanics are also outlined.

This book is intended for physicists, mathematicians, engineers, historians and philosophers of science. It is also addressed to teachers of physics at university or high school levels and to their students. After all, those who have taught and learned Newtonian mechanics know the difficulties and subtleties of its basic concepts (inertial frame of reference, fictitious centrifugal force, inertial and gravitational masses, etc.) Above all, it is intended for young unprejudiced people who have an interest in the fundamental questions of mechanics: Is there an absolute motion of any body relative to space or only relative motion between material bodies? Can we prove experimentally that a body is accelerated relative to space or only relative to other bodies? What is the meaning of inertia? Why do two bodies of different weight, form and chemical compositions fall with the same acceleration in vacuum on the earth's surface? When Newton rotated the bucket and saw the water rising towards the sides of the bucket, what was responsible for this effect? Was it due to the rotation of the water relative to some material body? What flattens the earth at the poles in its diurnal rotation? Is it the rotation of the earth relative to something? Is the earth really rotating and translating? We show that the answer to these questions with relational mechanics is much simpler and more philosophically sound and appealing than in Einstein's theories of relativity.

Nowadays the majority of physicists accept Einstein's theories as correct. We show this is untenable and present an alternative theory which is much clearer and more reasonable than the previous ones. We know that these are strong statements, but we are sure that anyone with a basic understanding of physics will accept this fact after reading this book with impartiality and without prejudice. With an understanding of relational mechanics, we enter a new world, viewing the same phenomena with different eyes and from a new perspective. It is a change of paradigm [1]. This new formulation will help put physics on new rational foundations, moving it away from the mystifications of this century.

We hope physicists, engineers, mathematicians and philosophers will adopt this book in their courses of mechanics, mathematical methods of physics and history of science, recommending it to their students. We believe the better way to create critical minds and to motivate the students is to present to them different approaches for the solution of the same problems, how the concepts have been growing and changing throughout history and how great scientists viewed equivalent subjects from different perspectives.

A Portuguese version of this book was published under the title *Mecânica Relacional*, [2].

In this book we utilize the International System of Units. When we define any physical concept we utilize "$\equiv$" as a symbol of definition. We utilize symbols with a double subscript with three different meanings. Examples: $\vec{F}_{ji}$ is the force exerted by particle j on particle i, $\vec{a}_{12} = \vec{a}_1 - \vec{a}_2$ is the acceleration of particle 1 minus the acceleration of particle 2, and $\vec{v}_{mS}$ is the velocity of particle m relative to the frame of reference S. In the text we clarify which meaning we are employing in each place.

Part I

Old World

Chapter 1

Newtonian Mechanics

1.1 Introduction

The branch of knowledge which deals with the equilibrium and motion of masses is called mechanics. For the last three hundred years the mechanics taught in schools and universities has been based on the work of Isaac Newton (1642-1727). His main book is called *Mathematical Principles of Natural Philosophy*, usually known by its first Latin name, *Principia*, [3]. Originally published in 1687, it is based on the concepts of space, time, velocity, acceleration, weight, mass, force, *etc.* In the next section we present Newton's own formulation of mechanics.

Since long before Newton, there has always been a great debate between philosophers and scientists regarding the distinction between absolute and relative motion. In other words, motion of a body relative to empty space and relative to other bodies. For a clear discussion of this whole subject with many quotations from the original see the authoritative book by Julian Barbour, *Absolute or Relative Motion?* [4]. In our book we consider only Newton and others following him. The reasons for this are the impressive success of his mechanics and the new standard he introduced in this whole discussion with his dynamical arguments, as distinguished from kinematical arguments, in favour of absolute motion. In particular, we can cite his famous bucket experiment. This is one of the main subjects of this work.

1.2 Newtonian Mechanics

The *Principia* begins with eight definitions, [3]. The first definition is "quantity of matter," which nowadays we call the inertial mass of a body. Newton defined it as product of the density and volume occupied by the body:

> *Definition I: The quantity of matter is the measure of the same,*
> *arising from its density and bulk conjointly.*

Thus air of a double density, in a double space, is quadruple in
quantity; in a triple space, sextuple in quantity. The same thing
is to be understood of snow, and fine dust or powders, that are
condensed by compression or liquefaction, and of all bodies that are
by any causes whatever differently condensed. I have no regard in
this place to a medium, if any such there is, that freely pervades
the interstices between the parts of bodies. It is this quantity that
I mean hereafter everywhere under the name of body or mass. And
the same is known by the weight of each body, for it is proportional
to the weight, as I have found by experiments on pendulums, very
accurately made, which shall be shown hereafter.

Designating the quantity of matter (the inertial mass) of a body m_i, its
density ρ and its volume V we would have:

$$m_i \equiv \rho V \ . \tag{1.1}$$

Later on we will present Mach's criticism of this definition. We will also
discuss in detail the proportionality between the mass and weight of bodies, as
well as Newton's experiments on this matter. For the moment it is important
to stress that with this proportionality Newton found a precise operational way
of determining the mass of any body, as he needed only to weight it.

Then Newton defines the quantity of motion as the quantity of matter times
the velocity of the body:

> *Definition II: The quantity of motion is the measure of the same,*
> *arising from the velocity and quantity of matter conjointly.*

The motion of the whole is the sum of the motions of all the parts;
and therefore in a body double in quantity, with equal velocity, the
motion is double; with twice the velocity, it is quadruple.

Denoting the vectorial velocity $\vec{v}$ and the quantity of motion $\vec{p}$ we have:

$$\vec{p} \equiv m_i \vec{v} \ .$$

Newton goes on to define the inertia of a body:

> *Definition III: The* vis insita, *or innate force of matter, is a power of resisting, by which every body, as much as in it lies, continues in its present state, whether it be of rest, or of moving uniformly forwards in a right line.*

> This force is always proportional to the body whose force it is and differs nothing from the inactivity of the mass, but in our manner of conceiving it. A body, from the inert nature of matter, is not without difficulty put out of its state of rest or motion. Upon which account, this *vis insita* may, by a most significant name, be called inertia (*vis inertiae*) or force of inactivity. (...)

His fourth definition is "impressed force," namely: *An impressed force is an action exerted upon a body, in order to change its state, either of rest, or of uniform motion in a right line.*

Then follow definitions of centripetal force, of the absolute quantity of a centripetal force, of the accelerative quantity of a centripetal force and the motive quantity of a centripetal force.

After these 8 definitions there is a *Scholium* with the definitions of absolute time, absolute space and absolute motion. It is worthwhile quoting its main parts:

> Hitherto I have laid down the definitions of such words as are less known, and explained the sense in which I would have them to be understood in the following discourse. I do not define time, space, place, and motion, as being well known to all. Only I must observe, that the common people conceive those quantities under no other notions but from the relation they bear to sensible objects. And thence arise certain prejudices, for the removing of which it will be convenient to distinguish them into absolute and relative, true and apparent, mathematical and common.

> I. Absolute, true, and mathematical time, of itself, and from its own nature, flows equably without relation to anything external, and by another name is called duration: relative, apparent, and common time, is some sensible and external (whether accurate or unequable) measure of duration by the means of motion, which is commonly used instead of true time; such as an hour, a day, a month, a year.

> II. Absolute space, in its own nature, without relation to anything external, remains always similar and immovable. Relative space is some movable dimension or measure of the absolute spaces; which our senses determine by its position to bodies; and which is com-

monly taken for immovable space; such is the dimension of a sub-terraneous, an aerial, or celestial space, determined by its position in respect of the earth. Absolute and relative space are the same in figure and magnitude; but they do not remain always numeri-cally the same. For if the earth, for instance, moves, a space of our air, which relatively and in respect of the earth remains always the same, will at one time be one part of the absolute space into which the air passes; at another time it will be another part of the same, and so, absolutely understood, it will be continually changed.

III. Place is a part of space which a body takes up, and is according to the space, either absolute or relative. (...)

IV. Absolute motion is the translation of a body from one absolute place into another; and relative motion, the translation from one relative place into another. (...)

Then come his three "Axioms, or Laws of Motion" and six corollaries, namely:

Law I: Every body continues in its state of rest, or of uniform motion in a right line, unless it is compelled to change that state by forces impressed upon it.

Law II: The change of motion is proportional to the motive force impressed; and is made in the direction of the right line in which that force is impressed.

Law III: To every action there is always opposed an equal reaction: or, the mutual actions of two bodies upon each other are always equal, and directed to contrary parts.

Corollary I: A body, acted on by two forces simultaneously, will describe the diagonal of a parallelogram in the same time as it would describe the sides by those forces separately.

(...)

Corollary V: The motions of bodies included in a given space are the same among themselves, whether that space is at rest, or moves uniformly forwards in a right line without any circular motion.

(...)

His first law is usually called the law of inertia.

His second law of motion might be written as:

$$\vec{F} = \frac{d\vec{p}}{dt} = \frac{d}{dt}(m_i \vec{v}) \ . \tag{1.2}$$

Here we have used $\vec{F}$ for the resultant force acting on the body. If the inertial mass m_i is a constant, then this law can be cast in the simple and well-known form

$$\vec{F} = m_i \vec{a} \, , \tag{1.3}$$

where $\vec{a} = d\vec{v}/dt$ is the acceleration of the body.

His third law is called the law of action and reaction. Denoting the force exerted by a body A on another body B by $\vec{F}_{AB}$, and the force exerted by B on A by $\vec{F}_{BA}$, the third law states that:

$$\vec{F}_{AB} = -\vec{F}_{BA} \, .$$

Whenever Newton utilized the third law, the forces between two bodies were always directed along the straight line joining them, as in the law of gravitation.

His first corollary is called the law of the parallelogram of forces.

His fifth corollary introduces the concept of *inertial frames* (frames which are at rest or which move with a constant velocity relative to absolute space).

In Section XII of Book I of the *Principia*, Newton proved two extremely important theorems related to the force exerted by a spherical shell on internal and external points, supposing forces which fall off as the inverse square of the distance (as is the case with Newton's gravitational law and Coulomb's electrostatic force):

Section XII: The attractive forces of spherical bodies.

> Proposition 70. Theorem 30: If to every point of a spherical surface there tend equal centripetal forces decreasing as the square of the distances from these points, I say, that a corpuscle placed within that surface will not be attracted by those forces any way.

If the body is anywhere inside the shell (not only on its center), it will not experience any resultant force from the shell as a whole.

> Proposition 71. Theorem 31: The same things supposed as above, I say, that a corpuscle placed without the spherical surface is attracted towards the centre of the sphere with a force inversely proportional to the square of its distance from that centre.

This means that a body outside the shell is attracted as if the shell were concentrated at its center.

In the third book of the *Principia* Newton presented his law of gravitation. This can be stated as follows: every particle of matter attracts every other particle with a force varying directly as the product of their gravitational masses and inversely as the square of the distance between them.

Nowhere in the *Principia* Newton did express the gravitational law in this form. But we can find statements similar to these in the following passages of the *Principia*: Book I, Props. 72 to 75 and Prop. 76, especially Corollaries I to IV; Book III, Props. 7 and 8; and in the General Scholium at the end of Book III. For instance, in Book I, Prop. 76, Cors. I to IV we read, referring to spheres with an isotropic distribution of matter, densities such as $\rho_1(r)$ and $\rho_2(r)$, in which every point attracts with a force which falls off as the square of the distance:

> Cor. I. Hence if many spheres of this kind, similar in all respects, attract each other, the accelerative attractions of each to each, at any equal distances of the centres, will be as the attracting spheres.

> Cor. II. And at any unequal distances, as the attracting spheres divided by the squares of the distances between the centres.

> Cor. III. The motive attractions, or the weights of the spheres towards one another, will be at equal distances of the centres conjointly as the attracting and attracted spheres; that is, as the products arising from multiplying the spheres into each other.

> Cor. IV. And at unequal distances directly as those products and inversely as the squares of the distances between the centres.

Proposition 7 of Book III states:

> *That there is a power of gravity pertaining to all bodies, proportional to the several quantities of matter which they contain.*

> That all planets gravitate one towards another, we have proved before; as well as that the force of gravity towards every one of them, considered apart, is inversely as the square of the distance of places from the centre of the planet. And thence (by Prop. 69, Book I, and its Corollaries) it follows that the gravity tending towards all the planets is proportional to the matter which they contain.

> Moreover, since all the parts of any planet A gravitate towards any other planet B; and the gravity of every part is to the gravity of the whole as the matter of the part to the matter of the whole; and (by Law III) to every action corresponds an equal reaction; therefore the planet B will, on the other hand, gravitate towards all the parts of the planet A; and its gravity towards any one part will be to the gravity towards the whole as the matter of the part to the matter of the whole. *Q. E. D.*

This last paragraph is very important. It shows the key role played by Newton's action and reaction law in the derivation of the fact that the force of gravity is proportional to the product of the masses of the two bodies (and not, for instance, proportional to the sum of the two masses, or to their product squared).

In the General Scholium at the end of the book we read:

> Hitherto we have explained the phenomena of the heavens and of our sea by the power of gravity, but we have not yet assigned the cause of this power. This is certain, that it must proceed from a cause that penetrates to the very centres of the sun and planets, without suffering the least diminution of its force; that operates not according to the quantity of the surfaces of the particles upon which it acts (as mechanical causes used to do), but according to the quantity of the solid matter which they contain, and propagates its virtue on all sides to immense distances, decreasing always as the inverse square of the distances.

In the *System of the World* written by Newton we can also see the importance of the law of action and reaction for the derivation of the fact that the gravitational force is proportional to the product of the masses. Here we quote Section 20 of the *Principia*, [3, p. 568], just after the Section where Newton discussed his pendulum experiments which showed the proportionality between weight and inertial mass:

> Since the action of the centripetal force upon the bodies attracted is, at equal distances, proportional to the quantities of matter in those bodies, reason requires that it should be also proportional to the quantity of matter in the body attracting.

> For all action is mutual, and (by the third Law of Motion) makes the bodies approach one to the other, and therefore must be the same in both bodies. It is true that we may consider one body as attracting, another as attracted; but this distinction is more mathematical than natural. The attraction resides really in each body towards the other, and is therefore of the same kind in both.

Algebraically his law of gravitation might be written as:

$$\vec{F}_{21} = -G\frac{m_{g1}m_{g2}}{r^2}\hat{r} \ . \tag{1.4}$$

In this equation $\vec{F}_{21}$ is the force exerted by the material particle 2 on the material particle 1, G is a constant of proportionality, m_{g1} and m_{g2} are the gravitational masses of particles 1 and 2, r is their distance and $\hat{r}$ is the unit vector pointing from 2 to 1.

Here we are calling the masses which appear in Eq. (1.4) "gravitational masses," to distinguish them from the "inertial masses" which appear in Newton's second law of motion, Eqs. (1.2) and (1.3). They might also be called "gravitational charges," by analogy with the electrical charges which appear in Coulomb's force, to be discussed later on. The electrical charges generate and experience electrical forces, while gravitational masses generate and experience gravitational forces. In this respect and observing the form of the force laws of universal gravitation and of Coulomb, the gravitational masses have a greater resemblance to electrical charges than inertial masses. Later on we discuss this in greater detail.

Utilizing Newton's law of gravitation, Eq. (1.4), and his theorems stated above, we find that a spherically symmetrical body will attract an external body as if all the gravitational mass of the spherical body were concentrated at its center. In the case of the earth, neglecting the small effects due to its form being not exactly spherical, this yields:

$$\vec{F} = -G\frac{M_{gt}m_g}{r^2}\hat{r} \ ,$$

where M_{gt} is the gravitational mass of the earth and $\hat{r}$ points radially outwards. This force is usually called the weight of the body, and is represented by $\vec{P}$:

$$\vec{P} = m_g\vec{g} \ , \tag{1.5}$$

where

$$\vec{g} = -\frac{GM_{gt}}{r^2}\hat{r} \ .$$

Here $\vec{g}$ is called the gravitational field of the earth. It is the downward acceleration of freely falling bodies, as we will see.

If we are close to the surface of the earth, then $r \approx R_t$, where R_t is the earth's radius. Near the surface of the earth the measured value of this acceleration is found to be: $g = |\vec{g}| \approx GM_{gt}/R_t^2 \approx 9.8$ m/s^2.

By performing experiments with pendulums, Newton established that the gravitational and inertial masses are proportional or equal to one another. He expressed this as a proportionality between matter (m_i) and weight ($m_g|\vec{g}|$) in Proposition 6 of Book III in the *Principia*:

> That all bodies gravitate towards every planet; and that the weights
> of bodies towards any one planet, at equal distances from the centre
> of the planet, are proportional to the quantities of matter which
> they severally contain.

Newton's Propositions 70 and 71 given above are presented nowadays as follows: We have a spherical shell of gravitational mass M_g and radius R centered on O, as in Figure 1.1. An element of mass dm_{g2} located at $\vec{r}_2$ in this spherical shell is given by $dm_{g2} = \sigma_{g2} da_2 = \sigma_{g2} R^2 d\Omega_2 = \sigma_{g2} R^2 \sin\theta_2 d\theta_2 d\varphi_2$, where $\sigma_{g2} = M_g/4\pi R^2$ is the uniform surface mass density, $d\Omega_2$ is the element of spherical angle, θ_2 and φ_2 are the usual angles of spherical coordinates, θ_2 ranging from 0 to π and φ_2 from 0 to 2π.

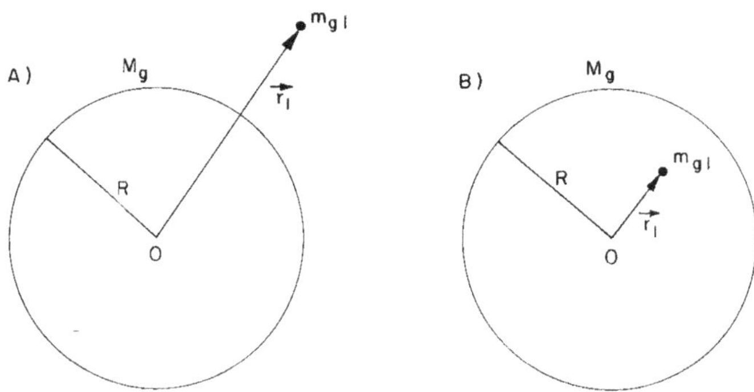

Figure 1.1: Spherical shell of mass M_g interacting with a mass point of mass m_{g1}.

The gravitational force exerted by this element of mass on the test particle m_{g1} located at $\vec{r}_1$ is given by Eq. (1.4), namely:

$$d\vec{F}_{21} = -G\frac{m_{g1}dm_{g2}}{r_{12}^2}\hat{r}_{12} \, ,$$

where $\vec{r}_{12} = \vec{r}_1 - \vec{r}_2$, $r_{12} = |\vec{r}_{12}|$ and $\hat{r}_{12} = \vec{r}_{12}/r_{12}$. Integrating this equation yields the following results, with $r_1 = |\vec{r}_1|$:

$$\vec{F} = \left\{ \begin{array}{ll} -GM_g m_{g1}\hat{r}_1/r_1^2 \, , & \text{if } r_1 > R \\ 0 \, , & \text{if } r_1 < R \, . \end{array} \right\} \tag{1.6}$$

If the test particle is outside the spherical shell it will be attracted as if the whole shell were concentrated at its center. If the test particle is anywhere inside the shell it will not experience any net gravitational force.

Nowadays, these theorems are easily proved utilizing Gauss's theorem. The force exerted by several masses on m_g located at $\vec{r}$ may be written as: $\vec{F} = m_g\vec{g}$, where $\vec{g}$ is the gravitational field at $\vec{r}$ due to the other masses. Gauss's theorem applied to the gravitational field states that the flux of $\vec{g}$ over a closed surface S is given by $-4\pi GM_{int}^g$, where M_{int}^g is the gravitational mass internal to S:

$$\oiint_S \vec{g} \cdot d\vec{a} = -4\pi G M^g_{int} \ .$$

$$(1.7)$$

Gauss's theorem is valid for any radial field which falls as $1/r^2$, as is the case for Newton's law. Let us calculate the gravitational field of a spherical shell of gravitational mass M_g and radius R centered on the origin O of a coordinate system. For reasons of symmetry, the gravitational field due to this spherical shell can only be radial, namely: $\vec{g} = g(r)\hat{r}$. We now consider a spherical surface S centered on O and with a radius $r > R$. The element of area of this spherical surface is $d\vec{a} = r^2 d\Omega \hat{r}$, where $d\Omega = \sin\theta d\theta d\varphi$ is the element of spherical angle. Utilizing Gauss's theorem we obtain:

$$g 4\pi r^2 = -4\pi G M_g \ ,$$

so that: $\vec{g}(r > R) = -G M_g \hat{r}/r^2$.

If we had integrated over a surface S such that $r < R$ than $M^g_{int} = 0$, so that we would arrive at: $\vec{g}(r < R) = 0$. With these results we recover Eq. (1.6).

Newton was completely aware of the cosmological implications of his 70th proposition, theorem 30 (the gravitational force on a test body anywhere inside a spherical shell is zero). The main implication is that we can essentially neglect the gravitational influence of the fixed stars on planetary motions and in experiments conducted on the earth, as the stars are randomly scattered in all directions in the sky (neglecting the concentration of stars in the Milky Way). He expressed this clearly in the second corollary of Proposion 14, Theorem 14 (The aphelions and nodes of the orbits of the planets are fixed), of Book III of the *Principia*:

> Cor. I. The fixed stars are immovable, seeing they keep the same position to the aphelion and nodes of the planets.

> Cor. II. And since these stars are liable to no sensible parallax from the annual motion of the earth, they can have no force, because of their immense distance, to produce any sensible effect in our system. Not to mention that the fixed stars, everywhere promiscuously dispersed in the heavens, by their contrary attractions destroy their mutual actions, by Prop. 70, Book I.

Newton discussed the distance of the fixed stars to the solar system at greater length in Section 57 of the *System of the World*, [3, pp. 596-7].

It is usually stated in textbooks that the gravitational constant G was measured by H. Cavendish (1731-1810) in 1798 with his torsion balance experiment. As a matter of fact, neither Newton nor Cavendish wrote the force law with G, as is given in Eq. (1.4), and they never mentioned the gravitational constant

G. Cavendish's paper is called "Experiments to determine the density of the earth," [5]. What he found is that the mean density of the earth is (5.448 ± 0.033) times greater than the density of water (Cavendish gave 5.48, due to an error in calculation corrected by A. S. Mackenzie, who reprinted Cavendish's work in 1899. See [5], Gravitation, Heat and X-Rays, pp. 100-101 and 143). For a discussion of his work see [6].

The claim that he measured G deserves an explanation. Considering the earth to be exactly spherical, the force it exerts on a material particle of gravitational mass m_g near its surface utilizing Eq. (1.4) is given by $P = GM_{gt}m_g/R_t^2 = m_g g$. The quantity of matter of the earth is given by its inertial mass $M_{it} = \rho_t \times V_t = \rho_t \times 4\pi R_t^3/3$, where ρ_t is its mean density, V_t its volume and R_t its radius. Newton found experimentally that the quantity of matter is proportional to the weight. Here we utilize this fact with a constant of proportionality equal to one, namely: $M_{it} = M_{gt} = M_t$. We then obtain $G = 3g/4\pi R_t \rho_t$. The gravitational field of the earth g near its surface has the same value as the acceleration of free fall. In the MKSA system of units we have: $g \approx 9.8$ m/s^2 and $R_t = 6.4 \times 10^6$ m. With Cavendish's measurement we get $\rho_t = 5.448 \times 10^3$ kg/m^3 and $M_t = 6 \times 10^{24}$ kg, where we have used the fact that the density of water is given by $\rho_{water} = 1$ g/cm^3 = 10^3 kg/m^3. This value of ρ_t in the previous expression for G yields: $G = 6.7 \times 10^{-8}$ cm^3/gs^2 = 6.7×10^{-11} m^3/kgs^2. The value given by modern tables is $G = 6.67 \times 10^{-11}$ m^3/kg s^2, which shows that Cavendish's measurement of the mean density of the earth is quite accurate.

We can then see that the value of G depends not only on the system of units but also on the choice of the constant of proportionality between the inertial and gravitational masses. If we had chosen $M_{it} = \alpha M_{gt}$, where the constant α could even have dimensions, then the value and dimensions of G would need to change to: $G = \alpha^2 \times (6.67 \times 10^{-11}$ m^3/kgs$^2)$. But this would not affect the results and predictions of any experiments. It is only a matter of convention to choose $\alpha = 1$ and this yields the usual value of G.

It should be remarked that Newton had a very good idea of the mean density of the earth 100 years before Cavendish. For instance, in Proposition 10 of Book III of the *Principia* he wrote:

> But that our globe of earth is of greater density than it would be if the whole consisted of water only, I thus make out. If the whole consisted of water only, whatever was of less density than water, because of its less specific gravity, would emerge and float above. And upon this account, if a globe of terrestrial matter, covered on all sides with water, was less dense than water, it would emerge somewhere; and, the subsiding water falling back, would be gathered to the opposite side. And such is the condition of our earth, which

in a great measure is covered with seas. The earth, if it was not
for its greater density, would emerge from the seas, and, according
to its degree of levity, would be raised more or less above their
surface, the water of the seas flowing backwards to the opposite
side. By the same argument, the spots of the sun, which float upon
the lucid matter thereof, are lighter than that matter; and, however
the planets have been formed while they were yet in fluid masses,
all the heavier matter subsided to the centre. Since, therefore, the
common matter of our earth on the surface thereof is about twice
as heavy as water, and a little lower, in mines, is found about three,
or four, or even five times heavier, it is probable that the quantity
of the whole matter of the earth may be five or six times greater
than if it consisted all of water; especially since I have before shown
that the earth is about four times more dense than Jupiter. (...)

Newton estimated $5\rho_{water} < \rho_t < 6\rho_{water}$ and Cavendish found 100 years
later $\rho_t = 5.5\rho_{water}$!

1.3 Energy

Newton based his mechanics in the concepts of force and acceleration. There
is another formulation based on the idea of energy. This formulation is due
originally to Huygens and Leibniz, although it has been later on incorporated
in newtonian mechanics. The basic concept is that of kinetic energy T. If we
are in an inertial frame S and a particle of inertial mass m_i moves in this frame
with a velocity $\vec{v}$ then its kinetic energy is defined by

$$T \equiv \frac{m_i v^2}{2} = m_i \frac{\vec{v} \cdot \vec{v}}{2} \; .$$

This kinetic energy is an energy of pure motion in classical mechanics. It is
not related to any kind of interaction (gravitational, electric, magnetic, elastic,
etc.) As such, it depends on the frame of reference, because the same body at
the same time may have different velocities relative to different inertial frames,
so that its kinetic energy relative to each one of these frames may have a different
value.

The other kinds of energy are based on how the particle interacts with
other bodies. For instance, the gravitational potential energy U_g between two
gravitational masses m_{g1} and m_{g2} separated by a distance r is given by

$$U_g = -G\frac{m_{g1}m_{g2}}{r} \; .$$

If the body m_{g1} is outside the earth at a distance r_1 from its center we can integrate this equation, replacing m_{g2} by dm_{g2} and assuming an isotropic matter distribution, to obtain

$$U = -G\frac{m_{g1}M_g}{r_1} \, ,$$

where M_g is the gravitational mass of the earth.

If the body is near the earth of radius R_t, at a distance h from its surface, $r_1 = R_t + h$, with $h \ll R_t$, this reduces to

$$U = -G\frac{m_{g1}M_g}{R_t + h} \approx m_{g1}gh - \frac{Gm_{g1}M_g}{R_t} \, ,$$

where $g = GM_g/R_t^2 \approx 9.8 \text{ m/s}^2$ is the gravitational field of the earth at its surface. Besides the constant term $-Gm_{g1}M_g/R_t$ this shows that the gravitational potential energy near the earth's surface is given by $m_{g1}gh$.

The analogous electrostatic potential energy U_e between two point charges q_1 and q_2 separated by a distance r is given by

$$U_e = \frac{1}{4\pi\varepsilon_o}\frac{q_1q_2}{r} \, ,$$

where $\varepsilon_o = 8.85 \times 10^{-12} \text{ C}^2 \text{ s}^2/\text{kg m}^3$ is the vacuum permittivity.

The potential elastic energy U_k of a mass interacting with a spring of elastic constant k is given by

$$U_k = \frac{kx^2}{2} \, ,$$

where x is the displacement of the body from the equilibrium position ($x = \ell - \ell_o$, with ℓ being the stretched length of the spring and ℓ_o its relaxed length).

We relate the concepts of force and energy by the equation

$$\vec{F} = -\nabla U \, . \tag{1.8}$$

This is especially useful when the potential energy and the force depend only on the positions of the bodies.

When we utilize the formulation of mechanics based only on the concept of energy, we utilize the theorem for the conservation of energy instead of Newton's three laws of motion. This law simply states that the total energy of the system (sum of the kinetic and potential energies) is a constant in time for conservative systems.

In this work we focus more on the Newtonian formulation based on forces.

Chapter 2

Applications of Newtonian Mechanics

Here we discuss several well-known applications of Newtonian mechanics. Later on we present Mach's criticisms of classical mechanics utilizing these examples. Lastly we present these examples from the point of view of relational mechanics to illustrate the different approach it makes possible.

In Newton's second law of motion, Eqs. (1.2) and (1.3), there appear a velocity and an acceleration (assuming a constant inertial mass). These velocities and accelerations are to be understood as referred to absolute space, and measured by absolute time. According to the fifth corollary we may also refer motion to any frame of reference which moves relative to absolute space with a constant velocity. Nowadays we call these frames of reference "inertial frames." In what follows, we assume that we are describing the motion of bodies in one inertial frame. Later on we will discuss this concept in more detail.

Here we consider only situations in which the inertial mass is a constant. In these cases Newton's second law of motion takes the form

$$\vec{F} = m_i \vec{a} \ . \tag{2.1}$$

Bodies with negligible dimensions compared with the distances involved in the problems are called particles. Usually we can neglect its internal properties and represent them by material points. A particle will be characterized by its mass, and for its localization we will utilize three coordinates describing its position: x, y, z. We are interested here in the motion of particles in paradigmatic situations.

2.1   Uniform Rectilinear Motion

If we have a particle which is free from external forces, or if the resultant force acting on this particle is zero, then the particle will move with a constant velocity $\vec{v}$ according to the first law of motion:

$$\vec{v} = \frac{d\vec{r}}{dt} = constant ,$$

$$\vec{r} = \vec{r}_o + \vec{v}t .$$

Here $\vec{r}(t)$ is the position vector of the body relative to an inertial system of reference, $\vec{r}_o$ the initial position of the particle and t the time. The velocity $\vec{v}$ is the velocity of the test body relative to an inertial frame of reference, or relative to Newton's absolute space.

The direction and magnitute of the velocity will be constant in time. This can only make sense if we know how to say when a particle is free from external forces (if we know in which conditions this happens). We also need to find an inertial system of reference without utilizing Newton's first law of motion (to avoid vicious circles). None of this is in any way simple or trivial.

2.2   Constant Force

We can easily integrate Eq. (2.1) when the force is a contant, yielding:

$$\vec{a} = \frac{d\vec{v}}{dt} = \frac{\vec{F}}{m_i} = constant ,$$

$$\vec{v} = \vec{v}_o + \vec{a}t ,$$

$$\vec{r} = \vec{r}_o + \vec{v}_o t + \frac{\vec{a}t^2}{2} .$$

Here $\vec{v}_o$ is the initial velocity.

2.2.1   Free Fall

As the first example of a constant applied force we have the free fall of a body near the surface of the earth, neglecting air resistance, Figure 2.1.

The only force acting on the test body is the gravitational attraction of the earth, namely, its weight $\vec{P} = m_g \vec{g}$. With Eq. (2.1) we get:

Figure 2.1: Free fall of a body of mass m.

$$\vec{a} = \frac{m_g}{m_i}\vec{g}\,.$$

The value of $\vec{g}$ depends only on the earth and on the location of the test body, but does not depend of m_i or m_g. The gravitational field $\vec{g}$ does not depend on the inertial or gravitational mass of the test body.

It is a fact of experience that all bodies fall in vacuum with the same acceleration near the surface of the earth. This fact cannot be derived from any of Newton's laws or mathematical theorems. This result is valid no matter what the weight, form or chemical composition of the bodies. If we have bodies 1 and 2 falling in vacuum at the same location near the earth's surface, we know from experience that $\vec{a}_1 = \vec{a}_2 = \vec{g}$, Figure 2.2. This means that $m_{g2}/m_{i2} = m_{g1}/m_{i1} = constant$.

The first to arrive at this conclusion was Galileo (1564-1642) when working with bodies falling on inclined planes. Some of the main results obtained by Galileo in mechanics date from the period 1600 to 1610. From these experiments (bodies falling on inclined planes with negligible resistance, or falling freely in vacuum) we obtain for all bodies:

$$\frac{m_i}{m_g} = constant = \alpha\,. \tag{2.2}$$

When we say that these two masses (inertial and gravitational) are equal ($\alpha = 1$), we are specifying $G = 6.67 \times 10^{-11}$ m^3/kg s^2. If we said that these two masses were proportional to one another, $m_i = \alpha m_g$ (where α might be a constant different from one and could even have dimensions), all the results would remain valid provided we had put $G = \alpha^2 \times (6.67 \times 10^{-11}$ m^3/kgs^2) instead of $G = 6.67 \times 10^{-11}$ m^3/kgs^2 in Newton's law of gravitation. To see this we need only write the acceleration of free fall as:

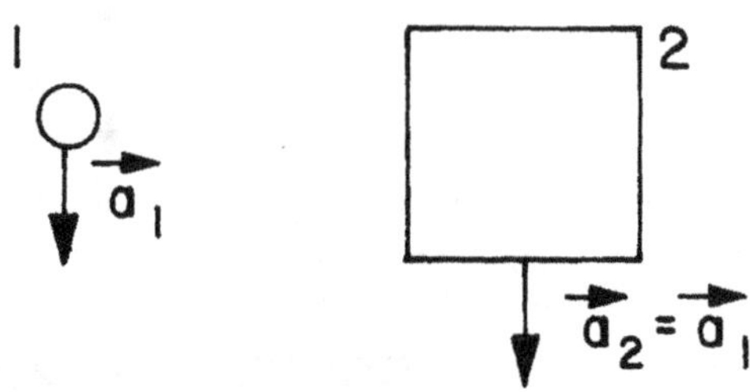

Figure 2.2: Two different bodies fall to the ground with the same acceleration in vacuum.

$$a = \frac{P}{m_i} = \frac{GM_{gt}m_g}{R_t^2 m_i} \ .$$

With Eq. (2.2) and $M_{it} = \rho_t V_t = \rho_t 4\pi R_t^3/3$ (from Newton's first definition) we get:

$$\frac{G}{\alpha^2} = \frac{3a}{4\pi R_t \rho_t} \ .$$

Putting the observed values of $a = 9.8$ m/s^2, $R_t = 6.4 \times 10^6$ m and $\rho_t = 5.5 \times 10^3$ kg/m^3 (from Cavendish's experiment) yields:

$$G = \alpha^2 \times (6.7 \times 10^{-11} \ m^3/kgs^2) \ .$$

The experiments of free fall only say that these two masses are proportional to one another and not that they are equal. As the choice of α has no influence in the predictions of experiments, it is simpler to say that they are equal to one another, choosing by convention $\alpha = 1$. From now on we will take this choice of α:

$$m_i = m_g \ . \tag{2.3}$$

The fact that in vacuum all bodies fall with the same acceleration was expressed as follows by Newton in the *Principia*: "It has been, now for a long time, observed by others, that all sorts of heavy bodies (allowance being made for the inequality of retardation which they suffer from a small power of resistance in the air) descend to the earth *from equal heights* in equal times; and that equality of times we may distinguish to a great accuracy, by the help of pendulums" (Book III, Proposition 6). In the *Opticks* he expressed it as follows: "(...) The open air in which we breathe is eight or nine hundred times lighter than water, and by consequence eight or nine hundred times rarer, and accordingly its resistance is less than that of water in the same proportion, or thereabouts, as I have found by experiments made with pendulums. And in thinner air the resistance is still less, and at length, by rarefying the air, becomes insensible. For small feathers falling in the open air meet with great resistance, but in a tall glass well emptied of air, they fall as fast as lead or gold, as I have seen tried several times" [7] (Book III, Query 28, p. 366).

2.2.2 Charge Moving Inside an Ideal Capacitor

We now present another example of a constant force. In 1784-5 Augustin Coulomb (1738-1806) obtained the law of force between two point charges q_1 and q_2. In modern vectorial notation and in the International System of Units the force exerted by q_2 on q_1 is given by:

$$\vec{F}_{21} = \frac{q_1 q_2}{4\pi\varepsilon_o} \frac{\hat{r}}{r^2} \, . \tag{2.4}$$

In this equation $\varepsilon_o = 8.85 \times 10^{-12}$ C^2 s^2/kg m^3 is the vacuum permittivity, r is the distance between the charges and $\hat{r}$ is the unit vector pointing from q_2 to q_1.

This force is very similar to Newton's law of gravitation, as it is directed along the straight line connecting the bodies, follows the law of action and reaction and falls as the inverse square of the distance. Moreover, it depends on the product of two charges, as in Newton's law it depends on the product of two masses. It would appear that Coulomb was led to this expression more by analogy with Newton's law of gravitation than by the results of his doubtful experiments [8]. The similarity between Coulomb's force (2.4) and Newton's law of gravity, Eq. (1.4), shows that the gravitational masses have the same role as the electrical charges: both generate and experience some kind of interaction with equivalent bodies, whether electrical or gravitational. The form of the interaction is essentially the same.

An ideal capacitor is represented in Figure 2.3. Two large square plates are separated by a distance $d \ll \ell$, where ℓ is the length of any plate. The plates situated at $z = z_o$ and $z = -z_o$ are uniformly charged with charges Q and $-Q$, respectively. In each plate we have a constant charge density given by $\sigma = Q/\ell^2$ and $-\sigma$, respectively. If we integrate the force exerted by the capacitor on an internal test charged particle utilizing Coulomb's force and neglecting edge effects we obtain the well known result:

$$\vec{F} = -q\frac{\sigma\hat{z}}{\varepsilon_o} = q\vec{E} \ . \tag{2.5}$$

Here $\hat{z}$ is the unit vector pointing from the negative to the positive plate and $\vec{E} = -\sigma\hat{z}/\varepsilon_o$ is the electric field generated by the capacitor in the region between the plates. Outside the capacitor there are no electric or magnetic fields.

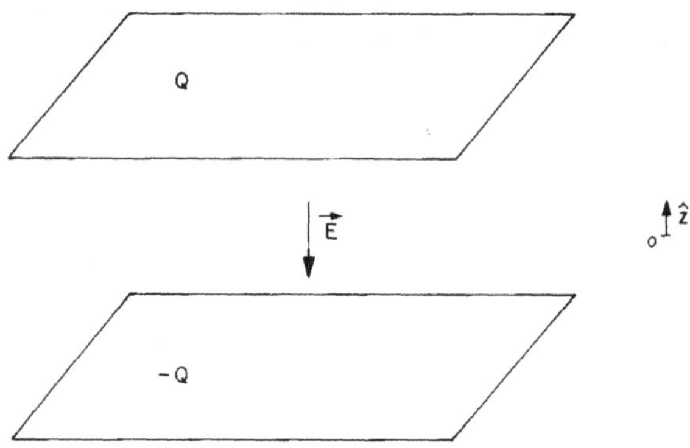

Figure 2.3: Ideal capacitor generating a uniform electric field between its plates.

In Weber's electrodynamics there will be a component of the force exerted by the capacitor on the test charge q moving inside it which depends on the velocity of q relative to the plates ([9], [10], [11, Section 5.6], [12, Sections 6.7 and 7.2], [13, Section 5.5], [14] and [15]). But supposing $v^2/c^2 \ll 1$, as we can consider in this experiment, Eq. (2.5) will also be valid in Weber's electrodynamics.

In classical electrodynamics (Maxwell's equations plus Lorentz's force) this is the total force exerted by the capacitor on the internal test charge, regardless of the velocity or acceleration of q relative to the plates, assuming fixed charges over the plates of the capacitor. This can be obtained by assuming a capacitor made of dielectric charged plates (with a vacuum between the plates) which do not allow a free motion of charges over its surface. Accordingly, the capacitor generates a constant electric field only between the plates, and no magnetic field.

Equating Eq. (2.5) with (2.1) yields:

$$\vec{a} = \frac{q}{m_i}\vec{E}\ .$$

The electric field depends only on the surface density of charge over the plates of the capacitor, and is independent of q or m_i. It is analogous to the gravitational field near the surface of the earth in our previous example. The difference now is that in the same electric field we can have bodies experiencing different accelerations. For instance, a proton (p) undergoes double the acceleration of an alpha (α) particle (nucleus of the helium atom, with two protons and two neutrons) if both are accelerated by the same capacitor: $\vec{a}_p = 2\vec{a}_\alpha$, as in Figure 2.4. This is due to the fact that the charge of an alpha particle is twice that of a proton, while its mass is four times that of the proton due to the two neutrons and two protons it contains. This does not happen in free fall, as all bodies, regardless of their weight, chemical composition, *etc.*, fall with the same acceleration in vacuum near the surface of the earth.

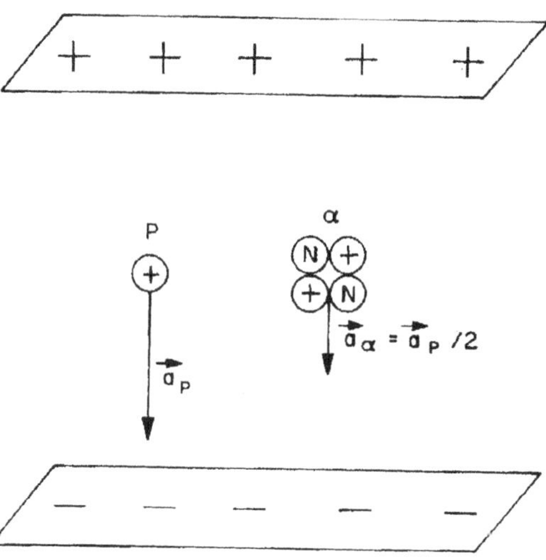

Figure 2.4: A proton and an alpha particle being accelerated inside a capacitor.

This is an extremely important fact. Comparing these two examples (see Figures 2.2 and 2.4) we can see that the inertial mass of a body is proportional to its gravitational mass, but not to its electrical charge. This fact suggests that the inertia of a body is related to its weight or gravitational property, but not to its electrical properties. Later on we will come back to this point.

2.2.3   Accelerated Train

The third example discussed here is an accelerated train moving along a straight line. From top of one of the wagons there is a small body suspended by a string, as in Figure 2.5.

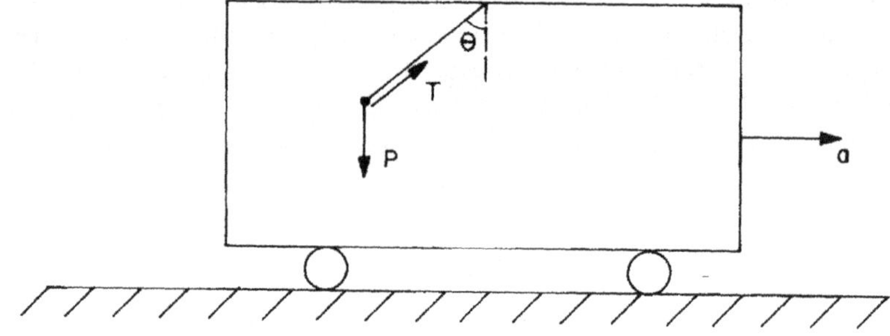

Figure 2.5: Accelerated train with a small body suspended by a string.

Here we analyse the equilibrium situation in which the body is at rest relative to the accelerated train. In other words, we analyse the situation when both of them have the same constant acceleration relative to the earth or to an inertial frame of reference. There are two forces acting on the body: the gravitational force of the earth (the weight $\vec{P} = m_g\vec{g}$), and the force exerted by the string due to its tension, $\vec{T}$. The equation of motion is

$$\vec{P} + \vec{T} = m_i\vec{a} \ .$$

Utilizing the angle θ of Figure 2.5:

$$P = T\cos\theta \ , \tag{2.6}$$

$$T\sin\theta = m_i a \ . \tag{2.7}$$

From these expressions and from $P = m_g g$ we obtain:

$$\tan\theta = \frac{m_i}{m_g}\frac{a}{g} \ . \tag{2.8}$$

From the experimental fact that θ is the same for all bodies independent of their weight, chemical composition *etc.* we obtain once more that $m_i = m_g$ or that the inertia of the body is proportional to its weight.

2.3 Oscillatory Motions

In this section we deal with forces which depend on position and which generate oscillatory motion.

2.3.1 Spring

The first example to be discussed here is that of a mass fastened to a spring which is connected to the earth, Figure 2.6. The weight of the test body is balanced by the normal force exerted by a frictionless table. The only remaining force is the horizontal force exerted by the spring.

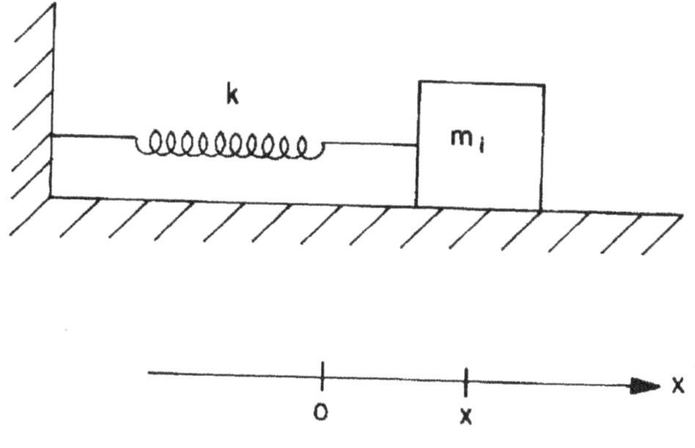

Figure 2.6: Spring on a frictionless table.

The force exerted by the spring on the body of inertial mass m_i is given by

$$\vec{F} = -kx\hat{x} \, , \tag{2.9}$$

where k is the elastic constant, x is the displacement of the body from the equilibrium position ($x = \ell - \ell_o$, with ℓ being the extended length of the spring and ℓ_o its relaxed length) and $\hat{x}$ the unit vector along the length of the spring.

Equating this with Eq. (2.1) with $\vec{a} = (d^2x/dt^2)\hat{x} = \ddot{x}\hat{x}$ yields the one-dimensional equation of motion:

$$m_i \ddot{x} + kx = 0 \; . \tag{2.10}$$

This equation can be easily solved:

$$x(t) = A \sin(\omega t + \theta_o) \; ,$$

where

$$\omega = \sqrt{\frac{k}{m_i}} \; . \tag{2.11}$$

The constant A is the amplitude of oscillation, θ_o is the initial phase and ω the frequency of oscillation. The constants A and θ_o may be related to the constant total energy E of the body and the initial position x_o by:

$$E = T + U = \frac{m_i \dot{x}^2}{2} + \frac{kx^2}{2} = \frac{kA^2}{2} \; ,$$

$$x_o = A \sin \theta_o \; .$$

2.3.2   Simple Pendulum

The second and most important example to be discussed here is a simple pendulum, Figure 2.7. A small body of typical dimension d oscillates in a vertical plane fastened to a string of constant length ℓ such that $d \ll \ell$.

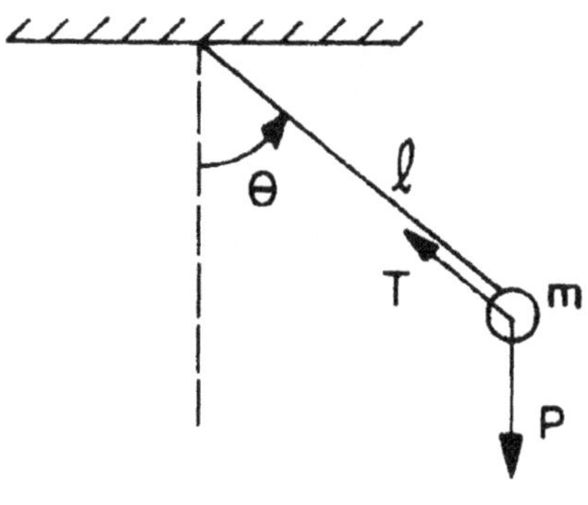

Figure 2.7: Simple pendulum of length ℓ.

Neglecting air resistance, there are two forces acting on the pendulum, its weight $\vec{P} = m_g \vec{g} = -m_g g \hat{z}$ and the tension in the string, $\vec{T}$. The equation of motion is simply $\vec{P} + \vec{T} = m_i\vec{a}$. Utilizing the angle represented in this Figure, the fact that the length of the string is a constant and a polar coordinate system (with $s = \ell\theta$, $v_\theta = \ell\dot\theta$ and $a_\theta = \ell\ddot\theta$ instead of x, $\dot{x}$ and $\ddot{x}$) yields

$$T - P\cos\theta = m\frac{v^2}{\ell} = m\ell\dot\theta^2 \ ,$$

$$-P\sin\theta = m_i a_\theta = m_i\ell\ddot\theta \ .$$

If we consider only small oscillations of the pendulum ($\theta \ll \pi/2$) then $\sin\theta \approx \theta$, and this last equation reduces to:

$$m_i\ddot\theta + m_g\frac{g}{\ell}\theta = 0 \ .$$

This equation has the same form as Eq. (2.10). Its solution is

$$\theta = A\cos(\omega t + B) \ , \tag{2.12}$$

with

$$\omega = \sqrt{\frac{m_g}{m_i}\frac{g}{\ell}} \ . \tag{2.13}$$

The constant A is the amplitude of oscillation for θ, B is the initial phase and ω the frequency of oscillation.

We now compare the frequencies of oscillation ω for the spring and for the simple pendulum, Eqs. (2.11) and (2.13). The periods of oscillation are given simply by $T = 2\pi/\omega$. The most striking difference is that while in the spring the frequency of oscillation depends only on m_i but not on m_g, in the pendulum the frequency of oscillation depends on the ratio m_g/m_i. Now suppose we have a test body of inertial mass m_i and gravitational mass m_g. If it is oscillating horizontally fastened to a spring of elastic constant k, its frequency of oscillation is given by $\omega_1 = \sqrt{k/m_i}$. If we connect two of these bodies to the same spring, the new frequency of oscillation is given by $\omega_2 = \sqrt{k/2m_i} = \omega_1/\sqrt{2}$, as in Figure 2.8.

On the other hand, if the first body were connected to a string of constant length ℓ and oscillating like a pendulum, its frequency of oscillation would be given by: $\omega_1 = \sqrt{m_g g/m_i\ell}$. Connecting two of these bodies to the same string, the new frequency of oscillation is given by $\omega_2 = \sqrt{2m_g g/2m_i\ell} = \omega_1$, as in Figure 2.9.

The same happens whatever the chemical composition of the test particle. In simple pendulums of the same length ℓ and at the same location on the

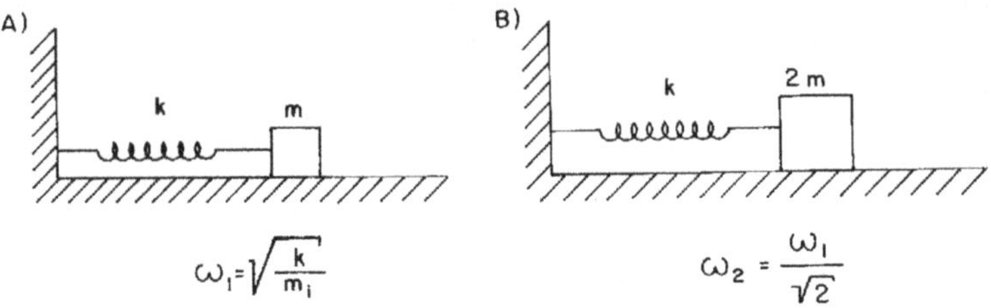

Figure 2.8: Two different masses attached to the same spring.

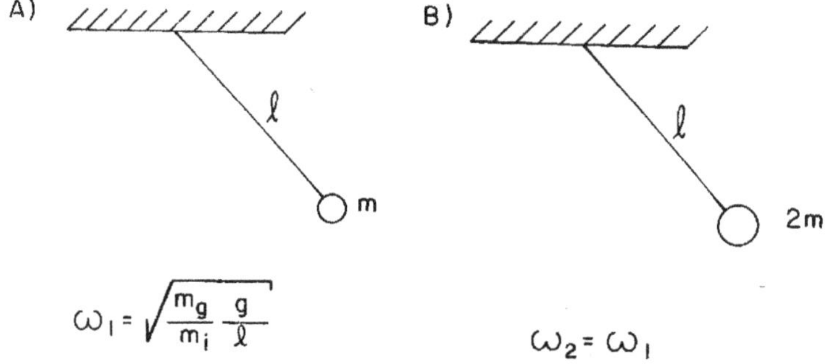

Figure 2.9: Two different masses attached to the same pendulum.

earth (same g), all bodies oscillate with the same frequency, regardless of their weight or chemical constitution, when air resistance is neglected. This is an experimental fact which cannot be derived from Newton's laws of motion (from Newton's laws we cannot derive that $m_i = m_g$ nor that $m_i/m_g = constant$). Only experience can tell us that the frequency of oscillation of a simple pendulum in vacuum does not depend on the weight or chemical constitution of the bodies, while the frequency of oscillation on an horizontal spring is inversely proportional to the square root of the mass of the body.

This experimental fact shows that we can cancel the masses in Eq. (2.13), writing the frequency of oscillation ω of the pendulum and its period T as:

$$\omega = \sqrt{\frac{g}{\ell}} = \frac{2\pi}{T} \, .$$

In section 2.2.2 we saw that the inertial mass of a body is proportional to the gravitational mass or weight of the body, but is not proportional to its charge or electrical properties. Here we see that the inertial mass of a body is not proportional to any elastic property of the body or of the surrounding medium (the spring in this case). Analogously, it can be shown that the inertial mass (or inertia) of a body is not related to the magnetic, nuclear, or any other property of the body or of the surrounding medium. Newton expressed this in Corollary V, Proposition 6 of Book III of the *Principia*, our words in square brackets: "The power of gravity is of a different nature from the power of magnetism; for the magnetic attraction is not as the matter attracted [the magnetic force is not proportional to the inertial mass of the attracted body]. Some bodies are attracted more by the magnet; others less; most bodies not at all. The power of magnetism in one and the same body may be increased and diminished; and is sometimes far stronger, for the quantity of matter, than the power of gravity; and in receding from the magnet decreases not as the square but almost as the cube of the distance, as nearly as I could judge from some rude observations."

The inertial mass is only proportional to the weight or gravitational mass of the body. Why does nature behave like this? There is no answer in Newtonian mechanics. We might imagine that a piece of gold could fall in vacuum with a larger acceleration than a piece of iron or silver of the same weight, but this is not the case. We might further imagine that a heavier lump of gold could fall in vacuum with a larger acceleration than a lighter lump of gold, or than another piece of gold with a different shape. Once more, this is not what happens. If any of these things did happen, all results of Newtonian mechanics might be kept, provided we did not cancel m_i with m_g. We would then conclude that m_i would depend on the chemical composition of the body, or on its form, or that it is not linearly proportional to m_g, or ..., depending on what were found experimentally.

Although this striking proportionality between inertia and weight does not prove anything, it is highly suggestive. It indicates that the inertia of a body (its resistance to acceleration) may have a gravitational origin. Later on, we show that this is indeed the case.

For the moment we present here Newton's own careful experiments with pendulums performed in order to arrive at this proportionality of inertia and weight (or proportionality between the quantity of matter m_i and m_g, as we would say today). In the first definition of the *Principia*, *quantity of matter*, Eq. (1.1), Newton wrote: "It is this quantity that I mean hereafter everywhere under the name of body or mass. And the same is known by the weight of each body, for it is proportional to the weight, as I have found by experiments on pendulums, very accurately made, which shall be shown hereafter." These experiments are contained in the previously mentioned Proposition 6, Theorem 6 of Book III of the *Principia*:

> *That all bodies gravitate towards every planet; and that the weights of bodies towards any one planet, at equal distances from the centre*

of the planet, are proportional to the quantities of matter which they severally contain.

It has been, now for a long time, observed by others, that all sorts of heavy bodies (allowance being made for the inequality of retardation which they suffer from a small power of resistance in the air) descend to the earth *from equal heights* in equal times; and that equality of times we may distinguish to a great accuracy, by the help of pendulums. I tried experiments with gold, silver, lead, glass, sand, common salt, wood, water, and wheat. I provided two wooden boxes, round and equal: I filled the one with wood, and suspended an equal weight of gold (as exactly as I could) in the centre of oscillation of the other. The boxes, hanging by equal threads of 11 feet, made a couple of pendulums perfectly equal in weight and figure, and equally receiving the resistance of the air. And, placing the one by the other, I observed them to play together forwards and backwards, for a long time, with equal vibrations. And therefore the quantity of matter in the gold (by Cor. I and VI, Prop. XXIV, Book II) was to the quantity of matter in the wood as the action of the motive force (or *vis motrix*) upon all the gold to the action of the same upon all the wood; that is, as the weight of the one to the weight of the other: and the like happened in the other bodies. By these experiments, in bodies of the same weight, I could manifestly have discovered a difference of matter less than the thousandth part of the whole, had any such been. (...)

From this experiment Newton found that $m_i = m_g$ within one part in a thousand:

$$\frac{m_i - m_g}{m_i} = \pm 10^{-3} \ .$$

With Eötvos's experiments at the turn of the century the precision of this relation improved to one part in 10^8. Nowadays it is known as one part in 10^{12}. For references, see [16].

2.3.3   Electrically Charged Pendulum

We now discuss the motion of a simple pendulum of length ℓ and inertial mass m_i performing small oscillations due to the gravitational attraction of the earth. Once more we suppose the earth to be a good inertial frame for this problem. The difference as regards subsection 2.3.2 is the following: Beyond its gravitational mass m_g, we suppose the pendulum to have an electrical charge q and to be in the presence of a permanent magnet, as in Figure 2.10.

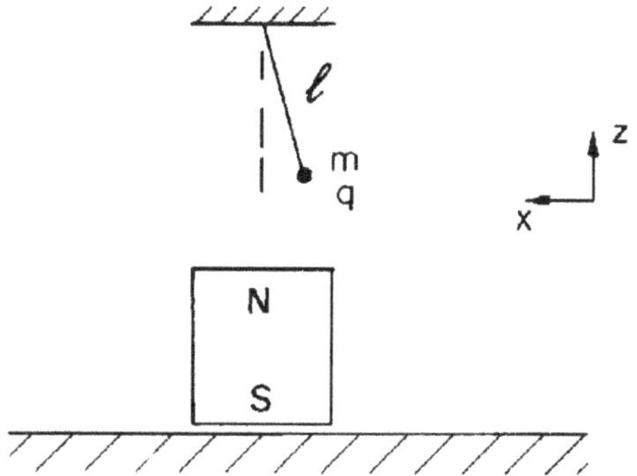

Figure 2.10: Charged simple pendulum oscillating near a magnet.

In this case the forces acting on the simple pendulum will be the gravitational force of the earth (the weight P), the tension T in the string and the magnetic force due to the magnet. In classical electromagnetism this force is represented by $\vec{F}_m = q\vec{v} \times \vec{B}$, where $\vec{v}$ is the velocity of the charge q relative to an inertial frame of reference (the earth or the laboratory in this case) and $\vec{B}$ is the magnetic field generated by the magnet. In Weber's electrodynamics the force exerted by the magnet on the charge has essentially the same value, although we don't necessarily need to speak of the magnetic field, and the velocity $\vec{v}$ will be the velocity of the charge relative to the magnet ([9], [17], [18], [19] and [12, Sections 6.7, 7.3 and 7.4]). As we are supposing the magnet to be at rest relative to the earth (assumed to be a good inertial frame in this experiment), there will not be any fundamental difference between the expressions for the magnetic force according to Weber's electrodynamics and classical electromagnetism. We then have a uniform gravitational field $\vec{g} = -g\hat{z}$ pointing downwards and a magnetic field $\vec{B} = B\hat{z}$ pointing vertically upwards. To simplify the analysis we will assume a uniform magnetic field (constant magnitude in space and time). The equation of motion takes the form

$$\vec{P} + \vec{T} + q\vec{v} \times \vec{B} = m_i \vec{a} \ . \tag{2.14}$$

We now suppose small oscillations ($\theta \ll \pi/2$) and that the pendulum is released from rest ($v_o = 0$) from the initial position $s_o = -|\theta_o|\ell \approx x_o$ with initial motion along the xz plane. With these conditions we find from (2.12) that in the absence of a magnetic field the velocity of the pendulum along the x axis is given approximately by

$$v_x \approx v_\theta = \ell\dot{\theta} = |\theta_o|\omega\ell \sin\omega t \ . \tag{2.15}$$

Here we put the initial phase equal to zero due to the initial conditions that the pendulum was released at rest from the initial angle θ_o. Moreover, we utilized $v_x \approx v_\theta$ because for small oscillations the motion is practically along the horizontal x axis. If there were no magnetic field, the pendulum would remain oscillating along the xz plane of our inertial frame of reference.

Supposing now the presence of the magnetic field, the motion of the pendulum will no longer remain along the same plane. With an initial velocity along the x axis, the vertical magnetic field will exert a force along the y axis given by

$$q\vec{v} \times \vec{B} = qv_x\hat{x} \times B\hat{z} = -qv_xB\hat{y} \ . \tag{2.16}$$

This force will modify the motion of the pendulum as indicated in Figure 2.11. In this Figure we are observing the projection of the motion of the pendulum in the yz plane as if we were on top of the pendulum. Assuming an initial motion along the positive x direction, $v_x > 0$, and a positive charge, $q > 0$, the magnet will deviate the motion of the pendulum in the direction $y < 0$. On the other hand, when the same pendulum is returning ($v_x < 0$) the magnet deviates it in the direction $y > 0$. This creates a clockwise rotation of the plane of oscillation of the pendulum with an angular velocity Ω (looking at the pendulum from above, supposing $q > 0$ and a magnetic field pointing vertically upwards).

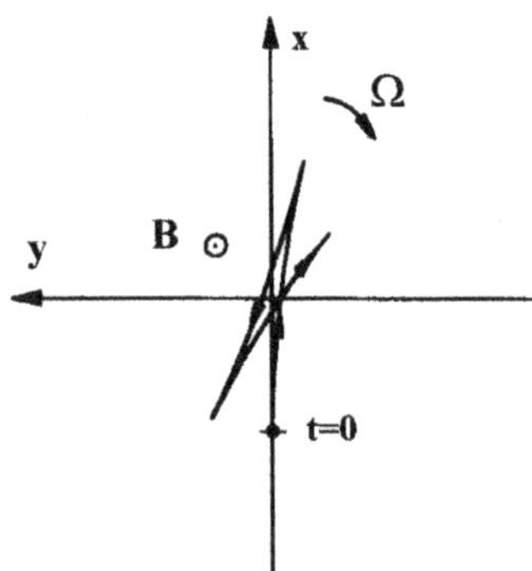

Figure 2.11: Rotation of the plane of oscillation of a charged pendulum due to a magnet.

We now calculate Ω assuming a weak magnetic field, namely, $qB/m_i\omega \ll 1$. This is analogous to having the greatest velocity in the x direction much larger than the greatest velocity in the y direction, or to saying that the velocity in the x direction is essentially unaffected by the magnet. From Eqs. (2.14), (2.15) and (2.16) the equation of motion in the y direction is given by (observing that $\vec{P} = -m_g g\hat{z}$ and that the tension $\vec{T}$ is in the xz plane):

$$-qv_x B = -q|\theta_o|\ell\omega \sin\omega t = m_i a_y \ .$$

(2.17)

This equation can be easily integrated twice utilizing that $v_y(t = 0) = 0$ and $y(t = 0) = 0$, yielding:

$$y = \frac{qB|\theta_o|\ell}{m_i}\left(\frac{\sin\omega t}{\omega} - t\right) \ .$$

(2.18)

The value of Ω can be obtained from Figure 2.12.

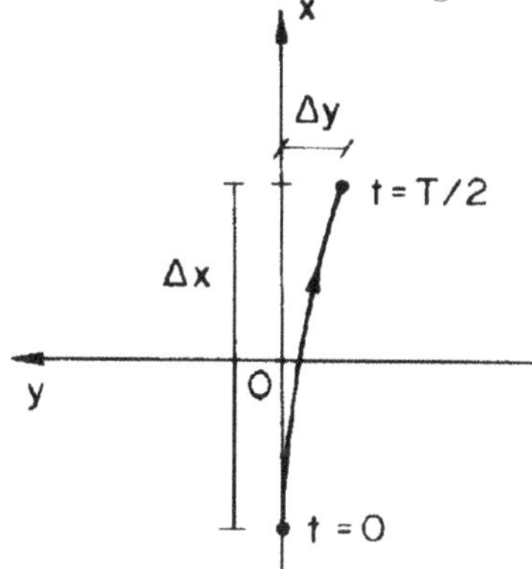

Figure 2.12: Geometry for calculating the precession of the plane of oscillation of a charged pendulum.

In half a period, $\Delta t = \pi/\omega$, the pendulum has moved from $x_o = -|\theta_o|\ell$ to $x = |\theta_o|\ell$, such that $\Delta x = 2|\theta_o|\ell$. On the other hand it has moved from $y_o = 0$ to $y(\pi/\omega) = -qB|\theta_o|\ell\pi/m_i\omega$, such that $\Delta y = -qB|\theta_o|\ell\pi/m_i\omega$. The value of Ω is then given by

$$\Omega = \frac{\Delta y/\Delta x}{\Delta t} = -\frac{qB}{2m_i} \ .$$

(2.19)

The negative value of Ω indicates a rotation in the clockwise direction when the pendulum is seen from above. To arrive at this result we neglected friction, and assumed uniform gravitational and magnetic fields, and that $qB/m_i\omega \ll 1$.

We conclude that the magnet causes a precession of the plane of oscillation of the charged pendulum oscillating in an inertial frame due to the action of a uniform gravitational force.

2.4   Uniform Circular Motion

In this section we discuss three situations of uniform circular motion which were analysed by Newton: A planet orbiting around the sun, two globes connected by a string and the spinning bucket.

We first consider a single body under the influence of a central force $\vec{F}$, shown in Figure 2.13.

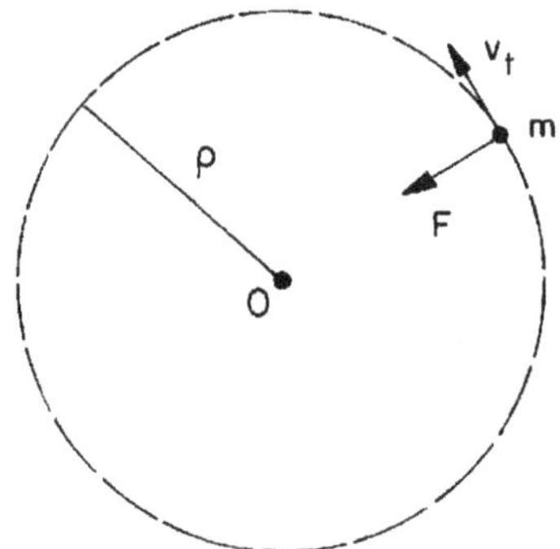

Figure 2.13: Uniform circular motion under a central force.

We consider a central force always directed toward the origin O of an inertial coordinate system S: $\vec{F} = -F\hat{\rho}$, where $F = |\vec{F}|$ and $\hat{\rho}$ is a unit vector pointing radially from O. With a polar coordinate system to describe the position, velocity and acceleration of a particle we have, respectively:

$$\vec{r} = \rho\hat{\rho} \; ,$$

$$\vec{v} = \frac{d\vec{r}}{dt} = \dot{\rho}\hat{\rho} + \rho\dot{\varphi}\hat{\varphi} \; ,$$

$$\vec{a} = \frac{d\vec{v}}{dt} = \frac{d^2\vec{r}}{dt^2} = (\ddot{\rho} - \rho\dot{\varphi}^2)\hat{\rho} + (\rho\ddot{\varphi} + 2\dot{\rho}\dot{\varphi})\hat{\varphi} \; ,$$

where $\rho = |\vec{r}|$.

With Eq. (2.1), and a constant ρ (uniform circular motion) we obtain:

$$F = m_i a_c = m_i \rho \dot{\varphi}^2 \ , \tag{2.20}$$

$$\dot{\varphi} = constant \ .$$

We represent the centripetal acceleration a_c arising from the motion in the φ direction by:

$$a_c = \rho \dot{\varphi}^2 = \frac{v_t^2}{\rho} \ ,$$

where

$$v_t = \rho \dot{\varphi} \ .$$

Huygens (1629-1695) and Newton were the first to obtain this value for the acceleration of a body orbiting with a constant velocity around a center. Huygens calculated the centrifugal force (a name created by him, meaning a tendency to depart from the center), arriving at his result in 1659. His manuscript on this topic was only published posthumously in 1703. However, in his book *Horologium Oscillatorum*, of 1673, he presented the main properties of the centrifugal force, but did not supply the proofs of how he arrived at these results. In any event he was the first to publish the correct value of this acceleration. Newton calculated the centripetal force (a name he framed later on in order to oppose to Huygens centrifugal force) between 1664 and 1666, without knowing Huygens's results. In the *Principia*, of 1687, he made great use of this result. See [4, Sections 9.7-9.8 and 10.5-10.6].

It should be observed that this central force changes only the direction of motion, leaving the magnitude of the tangential velocity constant: $|\vec{v}_t| = constant$.

2.4.1 Circular Orbit of a Planet

The first situation analysed here is a planet orbiting around the sun due to their mutual gravitational attraction. We consider the gravitational mass of the planet, m_{gp}, much smaller than the gravitational mass of the sun, m_{gs}, so that we can neglect the motion of the sun, see Figure 2.14. Although the orbit of the planets is in general elliptical, we consider here only the particular case of circular orbits in which the distances of the planets to the sun are constants in time.

From Eqs. (2.20) and (1.4) we obtain:

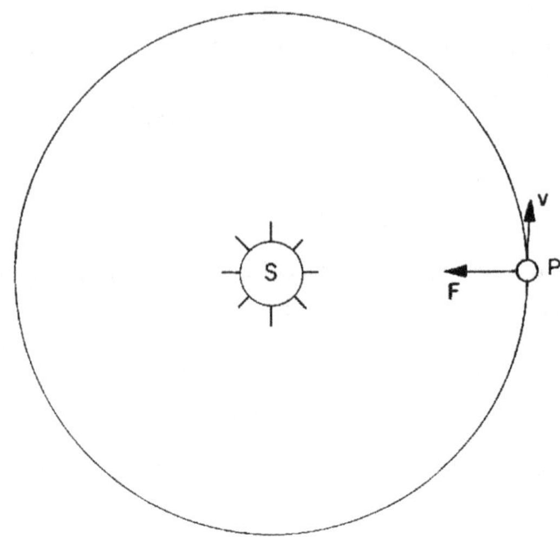

Figure 2.14: Planet orbiting around the sun.

$$F = G\frac{m_{gs}m_{gp}}{r^2} = m_{ip}a_{cp} = m_{ip}\frac{v_{tp}^2}{r} \ .$$

From Eq. (2.3):

$$a_{cp} = \frac{v_{tp}^2}{r} = \frac{Gm_{gs}}{r^2} \ .$$

The centripetal acceleration and the orbital velocity do not depend on the mass of the planet, but only on the mass of the sun.

How does the planet maintain a constant distance to the sun (or the moon to the earth, for instance) despite the gravitational attraction between them? According to Newton it is because the planet has an acceleration relative to absolute space (we might say relative to an inertial frame of reference). If the planet and the sun were initially both at rest relative to an inertial frame, they would attract and approach one another due to this attraction. What keeps the planet at a constant distance from the sun despite their gravitational attraction is the centripetal acceleration of the planet in absolute space (its tangential motion relative to the sun).

2.4.2 Two Globes

We now consider two globes connected to one another by a string and spinning relative to an inertial system with a constant angular velocity $\omega = \dot{\varphi} = v_t/\rho$ around the center of mass O, Figure 2.15.

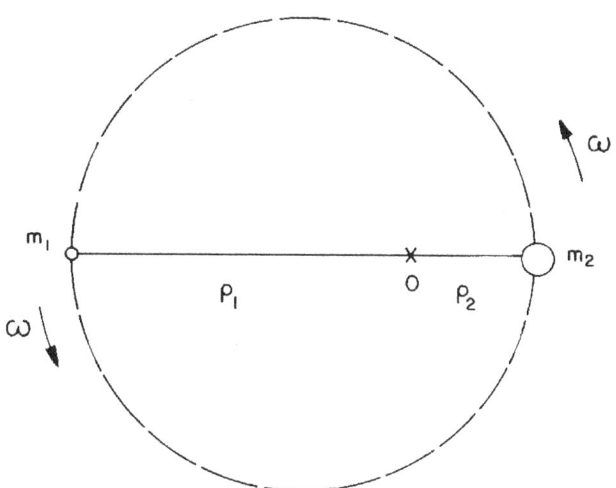

Figure 2.15: Two spinning globes connected by a string.

The only force exerted on each globe is due to the tension in the string. We call this tension T. By applying Eq. (2.20) to globe 1 we obtain:

$$T = m_{i1} a_{c1} = m_{i1} \frac{v_{t1}^2}{\rho_1} = m_{i1} w^2 \rho_1 \ , \tag{2.21}$$

For the second body we have, analogously:

$$T = m_{i2} a_{c2} = m_{i2} \frac{v_{t2}^2}{\rho_2} = m_{i2} w^2 \rho_2 \ . \tag{2.22}$$

The faster the rotation of the globes (*i.e.* the larger ω) the larger will be the tension in the string supporting the rotation. If instead of a string we had a spring of elastic constant k, the tension might be measured by $T = k(\ell - \ell_o)$, where ℓ is the stretched length of the spring ($\ell = \rho_1 + \rho_2$, see Figure 2.15) and ℓ_o its relaxed length. By measuring ℓ, given k and ℓ_o, we could know the tension. Knowing m_{i1}, ω and ρ_1 we can also obtain the tension T applying Eq. (2.21).

Newton discussed this problem of the two globes as a possible way of distinguishing the relative from absolute motion (or, more specifically, the relative rotation from the absolute rotation). By this experiment we could know if the globes were really rotating or not rotating relative to absolute space (or relative to an inertial frame). His discussion appears in the Scholium in the beginning of Book I of the *Principia*, following the first 8 definitions and before the three axioms or laws of motion. Here we present the entire discussion, with our emphasis:

> It is indeed a matter of great difficulty to discover, and effectually to distinguish, the true motions of particular bodies from the apparent; because the parts of that immovable space, in which those motions are performed, do by no means come under the observation of our senses. Yet the thing is not altogether desperate; for we have some arguments to guide us, partly from the apparent motions, which are the differences of the true motions; partly from the forces, which are the causes and effects of the true motions. For instance, if two globes, kept at a given distance one from the other by means of a cord that connects them, were revolved about their common centre of gravity, we might, from the tension of the cord, discover the endeavor of the globes to recede from the axis of their motion, and from thence we might compute the quantity of their circular motions. And then if any equal forces should be impressed at once on the alternate faces of the globes to augment or diminish their circular motions, from the increase or decrease of the tension of the cord, we might infer the increment or decrement of their motions; and thence would be found on what faces those forces ought to be impressed, that the motions of the globes might be most augmented; that is, we might discover their hindmost faces, or those which, in the circular motion, do follow. But the faces which follow being known, and consequently the opposite ones that precede, we should likewise know the determination of their motions. And thus we might find both the quantity and determination of this circular motion, even in an immense vacuum, where there was nothing external or sensible with which the globes could be compared. *But now, if in that space some remote bodies were placed that kept always a given position to one another, as the fixed stars do in our regions, we could not indeed determine from the relative translation of the globes among those bodies, whether the motion did belong to the globes or to the bodies. But if we observed the cord, and found that its tension was that very tension which the motions of the globes required, we might conclude the motion to be in the globes, and the bodies to be at rest;* and then, lastly, from the translation of the globes among the bodies, we should find the determination of their motions. But

how we are to obtain the true motions from their causes, effects, and apparent differences, and the converse, shall be explained more at large in the following treatise. For to this end was that I composed it.

Suppose the fixed stars to be at rest relative to absolute space. Spinning the globes with an angular velocity ω relative to absolute space (or relative to the fixed stars in this case) would, according to Newton, generate a tension in the string. This might be visualized by an increase in the length of a spring replacing the cord. Now suppose the same kinematical situation as above, namely, the globes rotating relative to the fixed stars with a constant angular velocity $\vec{\omega}$. But if in this second case the globes were at rest relative to absolute space and the fixed stars were revolving as a whole with an angular velocity $-\vec{\omega}$ relative to absolute space, then, according to Newton in this passage, there would be no tension in the cord (or the spring would not be streched or under tension). In this way we might distinguish the true or absolute rotation of the globes (relative to absolute space) from the apparent or relative rotation of the globes (relative to the fixed stars). Observing if there is or not a tension in the cord we might know if the globes were spinning or not relative to absolute space (or relative to an inertial frame of reference), although in both cases there would be the same relative or kinematical rotation of the globes relative to all other matter (the fixed stars here). Later on we will discuss this experiment further.

2.4.3 Newton's Bucket Experiment

We now analyse Newton's bucket experiment. This is one of the simplest and most important of all experiments performed by Newton. It is described just before the two-globes experiment presented above, in the Scholium following the eight definitions in the beginning of Book I of the *Principia*, before the presentation of the axioms or laws of motion (our emphasis):

> *The effects which distinguish absolute from relative motion are, the forces of receding from the axis of circular motion. For there are no such forces in a circular motion purely relative*, but in a true and absolute circular motion, they are greater or less, according to the quantity of motion. If a vessel, hung by a long cord, is so often turned about that the cord is strongly twisted, then filled with water, and held at rest together with the water; thereupon, by the sudden action of another force, it is whirled about the contrary way, and while the cord is untwisting itself, the vessel continues for some time in this motion; the surface of the water will at first be plain, as before

the vessel began to move; but after that, the vessel, by gradually communicating its motion to the water, will make it begin sensibly to revolve, and recede by little and little from the middle, and ascend to the sides of the vessel, forming itself into a concave figure (as I have experienced), and the swifter the motion becomes, the higher will the water rise, till at last, performing its revolutions in the same times with the vessel, it becomes relatively at rest in it. This ascent of the water shows its endeavor to recede from the axis of its motion; and the true and absolute circular motion of the water, which is here directly contrary to the relative, becomes known, and may be measured by this endeavor. At first, when the relative motion of the water in the vessel was greatest, it produced no endeavor to recede from the axis; the water showed no tendency to the circunference, nor any ascent towards the sides of the vessel, but remained of a plain surface, and therefore its true circular motion had not yet begun. But afterwards, when the relative motion of the water had decreased, the ascent thereof towards the sides of the vessel proved its endeavor to recede from the axis; and this endeavor showed the real circular motion of the water continually increasing, till it had acquired its greatest quantity, when the water rested relatively to the vessel. *And therefore this endeavor does not depend upon any translation of the water in respect of the ambient bodies, nor can true circular motion be defined by such translation.* There is only one real circular motion of any one revolving body, corresponding to only one power of endeavoring to recede from its axis of motion, as its proper and adequate effect; but relative motions, in one and the same body, are innumerable, according to the various relations it bears to external bodies, and, like other relations, are altogether destitute of any real effect, any otherwise than they may perhaps partake of that one only true motion. (...)

Let us obtain the form of the surface and the pressure anywhere within the spinning bucket. We consider the water to be an ideal homogeneous incompressible fluid of density $\rho = 1$ g/cm^3.

In the first situation the bucket and water are at rest relative to an inertial system, shown in Figure 2.16.

Then the surface of the water is flat and the pressure within the liquid increases as a function of the depth h according to $p(h) = p_o + \rho g h$, where $p_o = 1$ atm $= 760$ mm Hg $= 1 \times 10^5$ N/m^2 is the normal atmospheric pressure and $g \approx 9.8$ m/s^2 is the gravitational field of the earth. From this expression we may obtain Archimedes' (287-212 b. C.) rule: The upward force exerted by the water on any immersed body of volume V is given by the weight of the fluid displaced (in modern terms this is given by $\rho g V$). See *On Floating Bodies* in [20] and [21, pp. 538-560, especially Propositions 6 and 7]. This force does not depend on the mass of the body, but only on its immersed volume and the density of the surrounding liquid (or on the weight of the displaced fluid).

Now consider the bucket and water spinning together at a constant angular velocity ω relative to an inertial frame of reference (we may consider the earth as a good inertial system in this case). The water forms a concave figure, as represented in Figure 2.17.

Figure 2.16: Bucket and water at rest relative to the earth.

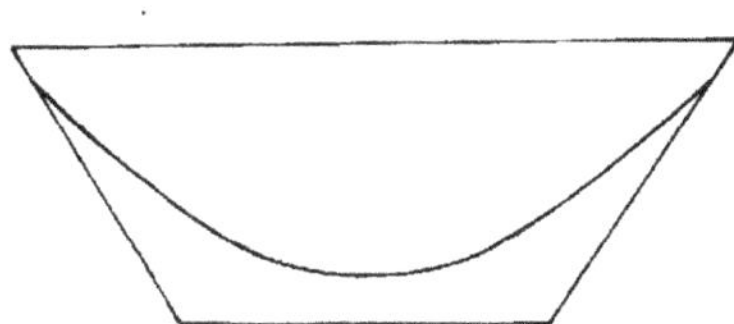

Figure 2.17: Bucket and water spinning together relative to the earth (Newton's bucket experiment).

The simplest way to obtain the form of the surface is to consider a frame of reference centered on the lower part of the spinning liquid with the z axis pointing vertically upwards, as in Figure 2.18.

Let us consider a small volume of liquid $dm_i = \rho dV$ just below the surface. It is acted upon by the downward force of gravity, $dP = dm_g g$, and by a force normal to the surface of the liquid due to the gradient of pressure, dE. As this portion of liquid moves in a circle centered on the z axis, there is no net vertical force. There is only a centripetal force pointing towards the z axis changing its direction of motion, but not the magnitude of the tangential velocity. From Figure 2.18 we obtain in this case (x being the distance of dm_i to the z axis):

$$dE \cos \alpha = dP = dm_g g \ , \tag{2.23}$$

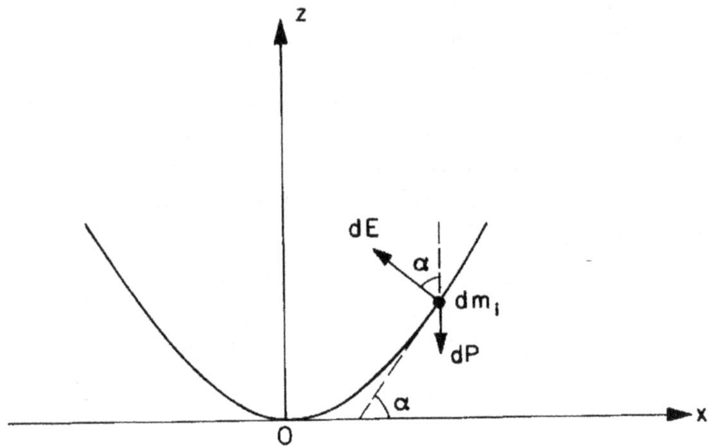

Figure 2.18: Geometry to calculate the form of the water surface when spinning.

$$dE \sin \alpha = dm_i a_c = dm_i \frac{v_t^2}{x} = dm_i \omega^2 x \ . \tag{2.24}$$

From these two equations and utilizing the fact that $dm_i = dm_g$ we have

$$\tan \alpha = \frac{\omega^2}{g} x \ . \tag{2.25}$$

Utilizing $\tan \alpha = dz/dx$, where dz/dx is the inclination of the curve at each point, and the fact that we want the equation of the curve which contains the origin $x = z = 0$ yields

$$z = \frac{\omega^2}{2g} x^2 \ . \tag{2.26}$$

The curve is a paraboloid of revolution. The greater the value of ω, the larger the concavity of the surface.

We can also calculate the pressure anywhere within the liquid by similar reasoning, with Figure 2.19. The equation of motion of a small quantity of water dm_i is $d\vec{P} + d\vec{E} = dm_i \vec{a}$, where $d\vec{E}$ is the force due to the gradient of pressure.

For the element of mass represented in Figure 2.19:

$$d\vec{E} = -(\nabla p)dV = -\left(\frac{\partial p}{\partial x}\hat{x} + \frac{\partial p}{\partial z}\hat{z} \right) dV \ .$$

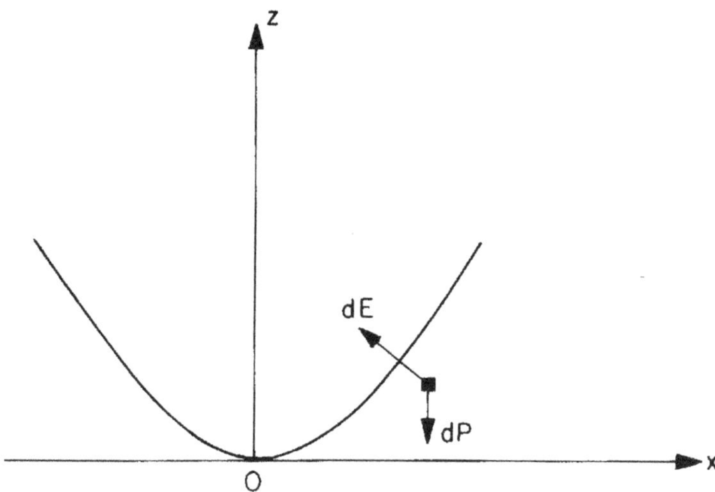

Figure 2.19: Forces in a volume element of water within the spinning bucket.

Utilizing the fact that there is only a centripetal acceleration yields $\vec{a} = -(v^2/x)\hat{x} = -\omega^2 x\hat{x}$. With $d\vec{P} = -dm_g g\hat{z}$ and $dm_i = dm_g$ we obtain

$$\frac{\partial p}{\partial z} = -\rho g \, ,$$

$$\frac{\partial p}{\partial x} = \rho\omega^2 x \, .$$

Integration of the first of these equations yields: $p = -\rho g z + f_1(x)$, where $f_1(x)$ is an arbitrary function of x. Integration of the second equation yields: $p = \rho\omega^2 x^2/2 + f_2(z)$, where $f_2(z)$ is another arbitrary function of z. Equating these two solutions and utilizing the fact that $p(x = 0, \ z = 0) = p_o$ yields the final solution (valid within the water):

$$p(x, \ z) = \frac{\rho\omega^2}{2}x^2 - \rho g z + p_o \, .$$

All over the surface of the liquid we have $p(x, \ z) = p_o$. Substituting this in the previous result once more yields the equation of the concave surface, namely, $z = \omega^2 x^2/2g$. This completes the solution of this problem in classical mechanics.

The importance of this experiment to Newton lies in the distinction it allows between absolute and relative rotation. According to Newton the surface will be concave only when the water is spinning relative to absolute space. This means that to him the ω which appears in Eq. (2.26) is the angular rotation of the water relative to absolute space and not the angular rotation of the water relative to any "ambient bodies." It is not the rotation of the water relative to the bucket, nor relative to the earth, nor even its rotation relative to the distant universe, such as the fixed stars. Remember that to Newton the absolute space has no relation to anything external, so that it is not related to the earth nor to the fixed stars.

We will now show that Newton had no other alternative at that time than to arrive at this conclusion. As the angular rotation of the bucket in Newton's experiment is much larger than the diurnal rotation of the earth or the annual rotation of the solar system, we may consider the earth to be without acceleration relative to the frame of fixed stars, and as a good inertial system during this experiment. In the first situation the bucket and the water are essentially at rest relative to the earth, so that they have at most a constant velocity relative to the fixed stars. The surface of the water is flat and there are no problems in deriving this conclusion. We now consider the second situation in which the bucket and the water are spinning together relative to the earth (and so relative to the fixed stars) with a constant angular velocity $\vec{\omega}_{be} = \vec{\omega}_{we} \equiv \vec{\omega} = \omega\hat{z}$. Here the z axis points vertically upwards from that location ($\hat{z} = \hat{r}$, where $\hat{r}$ points radially outwards from the earth's center), $\vec{\omega}_{be}$ is the angular velocity of the bucket relative to the earth and $\vec{\omega}_{we}$ is the angular velocity of the water relative to the earth. In this case the surface of the water is concave, rising towards the sides of the bucket. The key questions which need to be answered are: Why is the surface of water flat in the first situation and concave in the second? What is responsible for this different behaviour? The rotation of the water relative to what?

Let us analyse this from the Newtonian point of view. There are three main natural suspects for this concavity of the water: The rotation of the water relative to the bucket, relative to the earth, or relative to the fixed stars. That the bucket is not responsible for the different behaviour of the water can be immediately grasped by observing that there is no relative motion between the water and the bucket in both situations emphasized above. This means that whatever the force exerted by the bucket on each molecule of water in the first situation, it will remain the same in the second situation, as the bucket remains at rest relative to the water.

The second suspect is the rotation of the water relative to the earth. After all, in the first situation the water was at rest relative to the earth and its surface was flat, but when it was rotating relative to the earth in the second situation, its surface became concave. Thus, this relative rotation between the water and the earth might be responsible for the concavity of the water. Newton maintained that this is not the case ("And therefore this endeavor [to recede from the axis of circular motion] does not depend upon any translation of the water in respect of the ambient bodies, nor can true circular motion be defined by such translation"). We show here that Newton was consistent and correct in this conclusion when using his own law of gravitation. In the first situation, the only relevant force exerted by the earth on each molecule of water is of gravitational origin. As we saw in chapter 1, utilizing Eq. (1.4) and Newton's Theorem 31 we obtain that the earth attracts any molecule of water as if the earth were concentrated at its center, Eqs. (1.6) and (1.5): $\vec{P} = m_g \vec{g} = -m_g g \hat{z}$. In the second situation, the water is rotating relative to the earth, but the force exerted by the earth on each molecule of water is still given simply by $\vec{P} = -m_g g \hat{z}$ pointing vertically downwards. This is due to the fact that Newton's law of gravitation (1.4) does not depend on the velocity or acceleration between the interacting bodies. This means that in Newtonian mechanics, the earth cannot be responsible for the concavity of the surface of the water. Whether the water is at rest or spinning relative to the earth, it will experience the same gravitational force due to the earth, the weight $\vec{P}$ pointing downwards, without any tangential component of the force perpendicular to the z direction and depending on the velocity or acceleration of the water.

The third suspect is the set of fixed stars. In the first situation the water is essentially at rest or moving with a constant linear velocity relative to them and its surface is flat. In the second situation it is spinning relative to them and its surface is concave. This relative rotation might be responsible for the concavity of the water. But in Newtonian mechanics this is not the case either. The only relevant interaction of the water with the fixed stars is of gravitational origin. Let us analyse the influence of the stars in the first situation. As we saw in chapter 1, utilizing Eq. (1.4) and Newton's Theorem 30, we find that the net force exerted by all the fixed stars on any molecule of water is essentially zero, assuming that the fixed stars are distributed more or less at random in the sky and neglecting the small anisotropies in their distribution. This is the reason why the fixed stars are seldom mentioned in Newtonian mechanics. This will remain valid not only when the water is at rest relative to the fixed stars, but also when it is rotating relative to them. Once more, this is due to the fact that Newton's law of gravitation (1.4) does not depend on the velocity or acceleration between the bodies. Thus, his result (1.6) will remain valid no matter what the velocity or acceleration of body 1 relative to the spherical shell.

As we have seen, Newton was aware that we can neglect the gravitational influence of the set of all fixed stars in most situations. Recall what he wrote in the *Principia*: "(...) the fixed stars, everywhere promiscuously dispersed in the heavens, by their contrary attractions destroy their mutual actions, by Prop.

70, Book I." The conclusion is then that the relative rotation between the water and the fixed stars is not responsible for the concavity of the water either. Even introducing the external galaxies (which were not known by Newton) does not help, as they are known to be distributed more or less uniformly in the sky. So the same conclusion Newton reached for the fixed stars (that they exert no net force on other bodies) applies to the distant galaxies.

An important consequence of this fact is that even if the fixed stars and distant galaxies disappeared (were literally annihilated from the universe) or doubled in number and mass, the concavity of the water would not change in this experiment (according to Newtonian mechanics).

Since the effect of the concavity of the water is real and can be measured (the water can even pour out of the bucket), Newton had no other choice than to point out another cause for it, namely, the rotation of the water relative to absolute space. This was his only alternative, assuming the validity of his universal law of gravitation, which he proposed in the same book where he presented the bucket experiment. Moreover, this Newtonian absolute space cannot have any relation with the mass or quantitity of matter of the water, of the bucket, of the earth, of the fixed stars, the distant galaxies nor any other material body, as all these other possible influences have been eliminated.

A quantitative explanation of this key experiment without introducing absolute space is one of the main accomplishments of relational mechanics as developed in this book.

Chapter 3

Non-inertial Frames of Reference

We now discuss some of the examples of the previous chapter in non-inertial frames of reference S'. As we have seen, Newton's second law of motion is valid only in absolute space or in frames of reference which move with a constant translational velocity relative to it, by his fifth corollary. These are called inertial frames of reference which we represented by S. When the frame of reference is accelerated relative to an inertial frame, difficulties with the application of Newton's laws of motion arise. To overcome these difficulties it is necessary to introduce so-called fictitious forces. We analyse these situations here.

3.1 Constant Force

3.1.1 Free Fall

The first situation is the case of free fall. Suppose we are falling to the earth in vacuum. We will consider the earth an inertial frame of reference S. This means that our frame of reference S' is falling freely towards the earth with a constant acceleration, so that it is non-inertial. If we try to apply Newton's laws to study our own motion, we would write $F = m_i a'$ in order to find out our own acceleration a' in our frame of reference S'. The only force acting on us is the gravitational attraction of the earth, so that, $F = m_g g$. We would conclude that $a' = g \approx 9.8$ m/s^2 - which is wrong. After all we are at rest relative to ourselves, and the correct result we should have arrived at was $a' = 0$, Figure 3.1. In other words, we see ourselves at rest while the earth approaches us with an acceleration $\vec{a}_e = g\hat{z}$, where $\hat{z}$ points vertically upwards, from the earth to us.

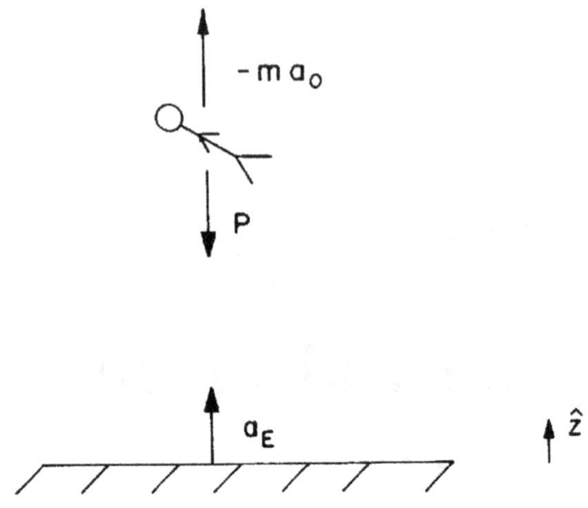

Figure 3.1: A person studying his own fall to the earth. The person is at rest relative to himself, while the earth moves towards him with an acceleration $a_E = g = 9.8$ m/s^2.

To arrive at the correct result we need to apply Newton's second law in the form

$$\vec{F} - m_i \vec{a}_o = m_i \vec{a}' \, , \tag{3.1}$$

where $\vec{a}_o$ is our acceleration relative to absolute space or relative to an inertial frame of reference, and $\vec{a}'$ is the acceleration of m_i relative to S'. In this case $\vec{a}_o = \vec{g} = -g\hat{z}$ (supposing the earth to be an inertial frame of reference, neglecting thus the small effects due to the non-inertial character of the earth's rotation). Utilizing this and the fact that $\vec{F} = m_g \vec{g}$, we would find that our acceleration relative to ourselves is given by (as always with $m_i = m_g$): $\vec{a}' = 0$. This is the correct answer in our own frame of reference.

The force $-m_i \vec{a}_o$ is called a fictitious force. The reason for this name is that all other forces which appear in $\vec{F}$ of Eq. (3.1) have a physical origin due to the interaction of the test body with other bodies, such as a gravitational interaction between the test body and the earth or the sun, an elastic interaction with a spring, an electric or magnetic interaction with another charge or magnet, a force of friction due to its interaction with a resistive medium, *etc.* On the other hand, the force $-m_i \vec{a}_o$ in classical mechanics has no physical origin. It is not due to an interaction of the test body with any other body. It only appears in non-inertial frames of reference which are accelerated relative to absolute space. At least this is the usual interpretation in Newtonian mechanics. Later on we will see that this need not be the case.

Despite its fictitious character, this force $-m_i \vec{a}_o$ is essential in non-inertial frames of reference in order to arrive at the correct results applying Newton's laws of motion.

3.1.2 Accelerated Train

The second example analysed here is an accelerated train. Once more we suppose the earth to be an inertial frame of reference as a good approximation here. In the previous chapter we analysed the motion and inclination of a pendulum in the frame of reference fixed to the earth. We now analyse the same problem in a frame of reference fixed in the accelerated wagon, as for instance, for a passenger who is inside the train, Figure 3.2. In this case the bob of mass m is at rest relative to the train and passenger, while the earth is accelerated to the left with an acceleration $\vec{a}_e = -a_e \hat{x}$.

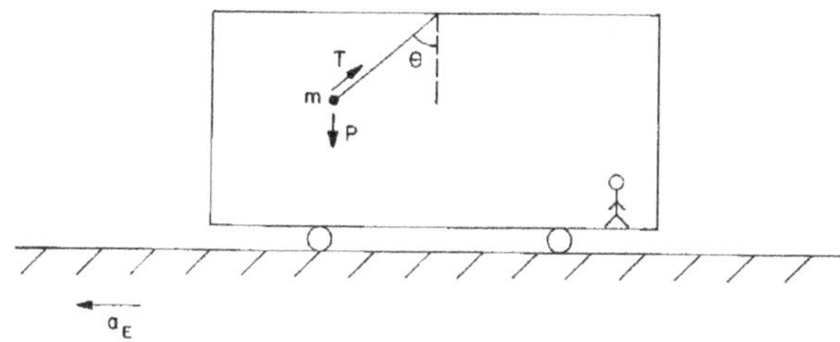

Figure 3.2: Passenger in an accelerated train. The mass m is at rest relative to him, while the earth moves to the left with an acceleration $\vec{a}_e = -|\vec{a}_e|\hat{x}$.

If the passenger applied Newton's second law to the bob of mass m in the form of Eq. (1.3) in the situation which he were observing, he would arrive at the same conclusion as Eqs. (2.6), (2.7) and (2.8), namely:

$$a' = g\frac{m_g}{m_i}\tan\theta \neq 0 .$$

But obviously this is the wrong answer in the frame of reference of the train. After all the pendulum is not moving relative to the train or to the passenger in the equilibrium situation being analysed here, so that the passenger should arrive at $a' = 0$. He can only obtain this with Eq. (3.1). He needs to introduce the fictitious force $-m_i \vec{a}_o$ in order to arrive at the correct result. In this case $\vec{a}_o = a_e \hat{x}$. This fictitious force balances the gravitational force exerted by the earth and the force exerted by the string, in order to cancel the motion of the pendulum relative to the passenger and keep it in an inclined position relative to the vertical, as represented in Figure 3.2. Then he would arrive at the situation described in Figure 3.3:

$$\vec{P} + \vec{T} - m_i \vec{a}_o = m_i \vec{a}' \ ,$$

so that $\vec{a}' = 0$.

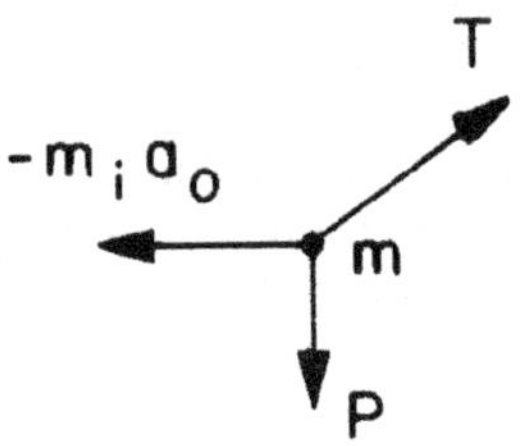

Figure 3.3: Forces in the passenger's frame of reference.

In this frame of reference the vertical component of the tension in the string balances the weight of the body, while $-m_i \vec{a}_o$ balances the horizontal component of the tension in the string, yielding zero motion of the bob relative to the train.

Once more in Newtonian mechanics there is no physical origin for this force $-m_i \vec{a}_o$, but it is essential to utilize it in the train's frame of reference to arrive at correct results.

3.2   Uniform Circular Motion

We now analyse some problems of the previous chapter in the frame of reference of the rotating bodies.

3.2.1   Circular Orbit of a Planet

We begin with the planet orbiting around the sun. Once more we consider here only the particular case of circular orbits. In the inertial frame of reference S considered previously, with the sun much more massive than the planet, the sun was considered essentially at rest and the planet was orbiting around the sun. Application of Eq. (1.3) yielded a centripetal acceleration given by $a_{cp} = G m_s / r^2$.

We now analyse this problem in the non-inertial frame of reference S' in which the sun and the planet are at rest. In other words, in a frame of reference

S' centered on the sun but that rotates together with the planet relative to S. In this frame we should find that the planet would not be accelerated, namely: $a' = 0$. But this is not the case if we apply Newton's second law of motion in the form of Eq. (1.3). In this new frame of reference, how can we explain the fact that the planet is at rest while subject to the gravitational attraction of the sun? How can the planet keep an essentially constant distance to the sun? To arrive at the correct result, *i.e.* that there is no acceleration of the planet in this frame of reference, and to explain why the distance between the planet and the sun is essentially constant, we need to introduce another fictitious force. In this case this fictitious force has a special name, centrifugal force, and is given by:

$$\vec{F}_c = -m_i\vec{\omega} \times (\vec{\omega} \times \vec{r}) \,, \tag{3.2}$$

where $\vec{r}$ is the position vector of the test body relative to the origin of the non-inertial system of reference and $\vec{\omega}$ is the vector angular velocity of the non-inertial system of reference relative to absolute space, or relative to any inertial frame of reference. In the previous chapter we considered the inertial frame of reference S centered on the sun. In this frame S the planet orbitated around the sun with an angular frequency ω. The non-inertial frame of reference S' considered here is also centered on the sun, but it rotates relative to S with the same angular frequency as the planet's orbit, shown in Figure 3.4.

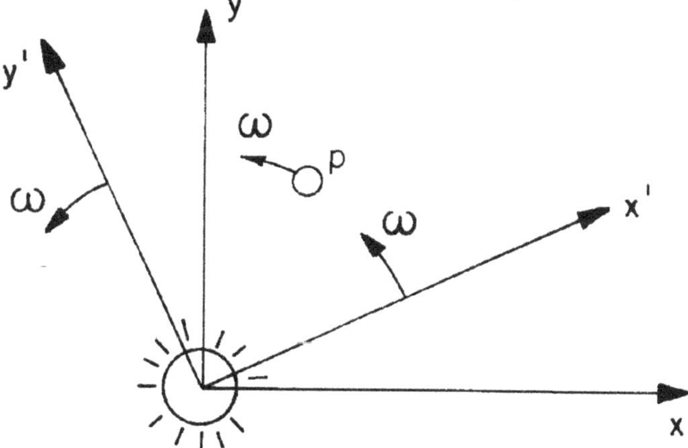

Figure 3.4: Frame S' rotating together with the planet relative to S.

If the planet were the earth, the period of rotation of S' relative to S would be $T = 2\pi/\omega = 365$ days. If the non-inertial frame of reference S' is rotating relative to the inertial frame S around the vertical z axis, $\vec{\omega} = \omega\hat{z}$. Then $\vec{\omega} \times (\vec{\omega} \times \vec{r}) = -\omega^2\rho\hat{\rho}$, where ρ is the distance of the test body to the axis of rotation and $\hat{\rho}$ is the unit vector pointing from the axis of rotation to the body, in a plane orthogonal to the axis of rotation (in polar coordinates: $\vec{r} = \rho\hat{\rho} + z\hat{z}$). In this case the centrifugal force is given simply by $\vec{F}_c = m_i\omega^2\rho\hat{\rho}$. This shows that this fictitious force, which appears only in the non-inertial frame of reference S' but not in S, is directed away from the center. This is the reason for the name "centrifugal," the case represented by Figure 3.5.

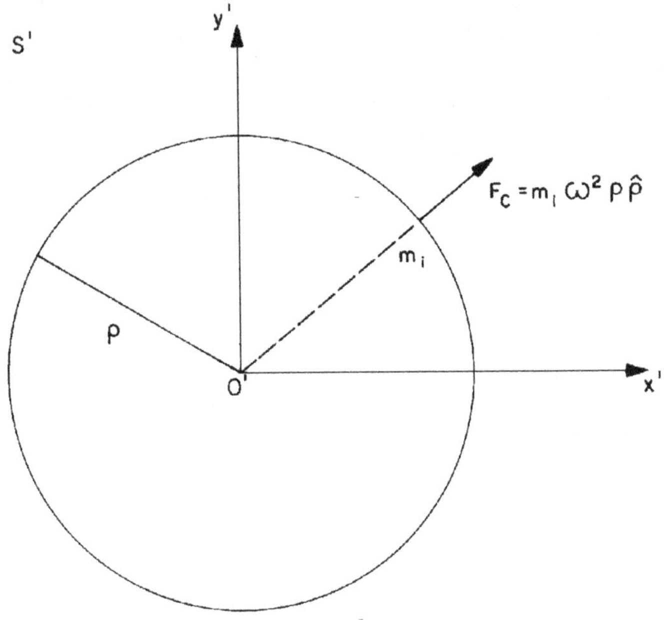

Figure 3.5: centrifugal force in the frame S'.

In this rotating non-inertial frame of reference Newton's second law of motion should be written as (in order to predict correct results):

$$\vec{F} + \vec{F}_c = m_i\vec{a}' \; , \tag{3.3}$$

where $\vec{F}$ is the resultant force due to all other bodies acting on m_i, $\vec{F}_c$ is given by Eq. (3.2) and $\vec{a}'$ is the acceleration of m_i relative to this non-inertial frame of reference.

In the problem of the planet we have the situation of Figure 3.6.

Utilizing the fact that $\vec{a}' = 0$ in this frame of reference gives the centrifugal force, namely

$$\vec{F}_c = G\frac{m_{gs}m_{gp}}{r^2}\hat{r} = -m_{ip}\vec{\omega} \times (\vec{\omega} \times \vec{r}) \; .$$

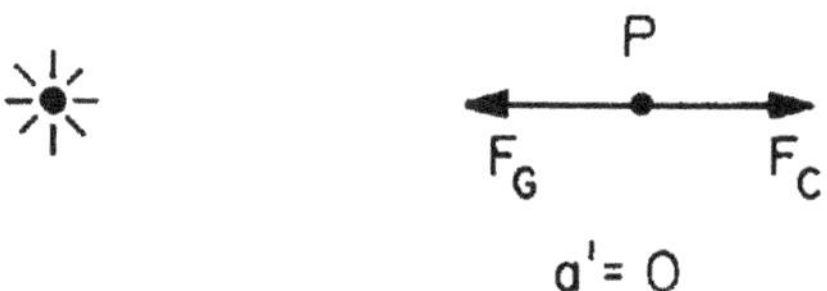

Figure 3.6: Planet "orbiting" around the sun, as seen in S'.

This yields: $\omega = \sqrt{Gm_{gs}/r^3}$. Alternatively we might use the fact that $\omega = \sqrt{Gm_{gs}/r^3}$ to find that $\vec{a}' = 0$ in the frame of reference S' in which the planet and the sun are at rest.

Once more there is no physical origin for this centrifugal force, while the gravitational force in this case is due to the attraction between the sun and the planet.

3.2.2 Two Globes

We now briefly discuss the experiment with two globes described by Newton. In a frame of reference S' which rotates with the globes and centered on the center of mass of the system, we have the situation depicted in Figure 3.7.

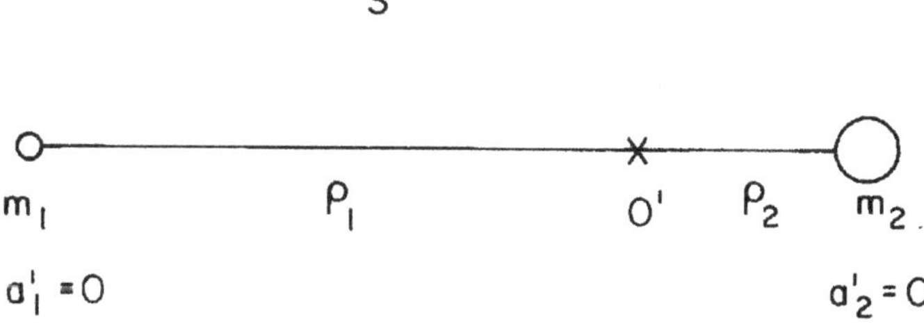

Figure 3.7: Two globes in the frame S'.

In this frame there is no motion of the globes despite the tension T in the string. The centripetal force due to this tension is balanced by a centrifugal force given by $m_i\omega^2\rho$, Figure 3.8: $F_{c1} = m_1\omega^2\rho_1 = T$ and $F_{c2} = m_2\omega^2\rho_2 = T$.

There are two interpretations of this equilibrium: Either we say that the tension is balanced by the centrifugal force, which does not let the globes ap-

Figure 3.8: Tension balanced by centrifugal force in the frame S'.

proach one another, or we say that the centrifugal force generates the tension in the string.

We might easily apply the same analysis to the previous problem of the sun and the planet, generalizing it to take into account the motion of the sun, and replacing the tension T in this example by the gravitational attraction $Gm_{g1}m_{g2}/r^2$. In this more real situation, the sun and the planet would be orbiting around the center of mass relative to an inertial frame of reference. In a non-inertial frame of reference centered on the center of mass and in which both bodies are at rest, the gravitational attraction would be balanced by the centrifugal force.

3.2.3   Newton's Bucket Experiment

We now consider the bucket experiment. We will concentrate on the situation in which the bucket and water rotate together with a constant ω relative to an inertial frame of reference S. In a frame of reference S' which rotates with the bucket there is no motion of the water, so that Eq. (3.3) reduces to (with Eq. (3.2) and $\vec{\omega} = \omega\hat{z}$):

$$-(\nabla p)dV - dm_g g\hat{z} + dm_i\omega^2\rho\hat{\rho} = dm_i\vec{a}' = 0 \ .$$

This yields the same result obtained previously, remembering that $dm_i = dm_g$ and that here we are utilizing $\rho\hat{\rho}$ instead of $x\hat{x}$ to represent the distance to the axis of rotation.

What is important to stress here and in the previous examples of the circular orbit of the planets and of the two globes, is that this centrifugal force has no physical origin in Newtonian mechanics. It appears in non-inertial frames of reference, and in this sense we might say that they are real (balancing the gravitational attraction of the sun, creating a tension in the string, pushing the water towards the sides of the bucket, *etc.*) On the other hand, unlike real forces such as the gravitational attraction exerted by the sun or by the earth, the electric force exerted by charges, the magnetic force exerted by magnets or current-carrying wires, or the elastic force exerted by a streched spring or tensioned cord, we cannot locate the material body responsible for the centrifugal force or for the ficitious forces in general. Let us analyse this in the case of the bucket experiment (a similar analysis can be carried out for all other examples discussed here).

We consider the situation in which the bucket and the water are rotating together relative to the earth and the fixed stars with a constant angular velocity around the vertical axis. We analyse the problem in the non-inertial frame of reference of the bucket, so that in this frame the surface of the water is concave, although the water is at rest. Is the bucket responsible for this concavity? No, after all the bucket is at rest relative to the water. Is the earth responsible for this concavity? In other words, is the rotation of the earth relative to the water, to the bucket and to this frame responsible for the centrifugal force? Once more the answer in Newtonian mechanics is no. As we saw in chapter 1, the gravitational force exerted by a spherical shell on material particles outside it points towards the center of the shell. As Newton's law of gravitation does not depend on the velocity or acceleration between the bodies, this will remain valid when the spherical shell is spinning. This means that according to Newtonian theory even when the earth is spinning relative to a material point or to a frame of reference, it will exert only the usual downward force of gravity, without any tangential force orthogonal to the radial direction. Are the fixed stars and external galaxies responsible for this concavity? In other words, is the rotation of the fixed stars (or of the external galaxies) relative to the water, to the bucket and to this non-inertial frame of reference responsible for the centrifugal force? The answer is no once more, due to Newton's 30th Theorem stated above, and to result (1.6). In other words, spherically symmetric distributions of matter do not exert any net gravitational forces on any internal point particles, regardless of the rotation or motion of these spherical distributions relative to the internal particle or to any frame of reference. This means that in Newtonian mechanics the fixed stars and distant galaxies might disappear without having any influence on the concavity of the water.

As we will see, relational mechanics will give a different answer here.

3.3 Rotation of the Earth

There are two main ways of determining the rotation of the earth. The first is kinematical or visual and the second, dynamical. We discuss the problem of the earth's rotation in this section.

3.3.1 Kinematical Rotation of the Earth

The simplest way to know that the earth rotates relative to something is by the observation of the astronomical bodies. Standing on the ground we do not observe the rotation of the earth directly; after all, we are essentially at rest relative to it. But looking at the sun we see that it moves around the earth with a period of one day. There are two obvious interpretations for this fact: The earth is at rest (as in the Ptolemaic system) and the sun translates around the earth; or the sun is at rest and it is the earth that spins around its axis (as in the Copernican system). Both interpretations are represented in Figure 3.9, considering the reference represented by the paper to be the rest frame.

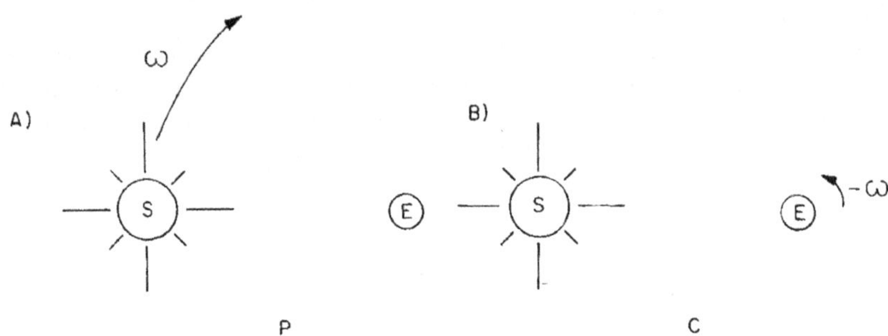

Figure 3.9: Relative rotation between the earth and the sun in the Ptolemaic (P) and Copernican (C) world views.

We can add to the motions of the earth and sun a common motion (a translation or rotation relative to absolute space, for instance) without altering their relative motion. What is important to realise here is that from the observed relative rotation between earth and sun we cannot determine which one of them is really moving relative to absolute space. The only thing observed and measured in this case is their relative motion. In this regard the Ptolemaic and Copernican systems are equally reasonable and compatible with the observations. It is a matter of taste to choose one or the other, considering only the relative rotation between the earth and the sun.

Another kinematical rotation of the earth is rotation relative to the fixed stars (relative to the stars which belong to our galaxy, the Milky Way). Although the moon, the sun, the planets and comets move relative to the background of stars, there is essentially no motion of one star relative to the others. The sky seen today with its constellations is essentially the same sky seen by the ancient Greeks or Egyptians. Although the set of stars rotate relative to the earth, they essentially do not move relative to one another and for this reason they are usually called fixed stars. Although the stellar parallax had been predicted by Aristarchus of Samos around 200 B.C., the first observation of this parallax (motion or change of position of one star relative to the others) was only obtained without doubt by F. W. Bessel in 1838. If we take a picture of the night sky with a long exposure we observe in the Northern hemisphere that all the stars rotate approximately around the north pole star, typically with a period of one day.

Once more, we may say that the real rotation belongs to the stars or to the spinning earth. We cannot decide between these two interpretations based only on these observations. It may be simpler to describe motions and planetary orbits in the frame of reference fixed with the stars than in the earth's frame of reference, but both of them are equally reasonable.

With a period of rotation of one day we get $\omega_k = 2\pi/T = 7 \times 10^{-5}$ rad/s, where ω_k is the kinematical rotation of the earth. The direction of this kinematical rotation is approximately the direction of the north pole star. In this way we have a complete description of the kinematical rotation, *i.e.*, the rotation of the earth relative to the fixed stars.

To simplify the analysis we are not distinguishing here the solar day to the sidereal day (time for the fixed stars to give a complete turn around the earth). We are putting both of them as 24 hours. As a matter of fact, while the sidereal day is essentially constant (when compared, for instance, with a mechanical or atomic clock), the solar day varies according to the month of the year. This was known to the Ancients, and Ptolemy (100-170 A.D.), for instance, presented the so called "equation of time" describing the variation of the solar day compared with the sidereal one. The mean solar day (obtained by an average taken over the year of the duration of the solar days) has by definition 24 hours, while the measured sidereal day has 23 hours, 56 minutes and 4 seconds. In a year the sun turns essentially 365 times around the earth, while the fixed stars turn 366 times. Another difference between these two motions is that while the stars rise at the same place relative to the earth's horizon all year long, the same does not happen with the sun, which rises at different locations at different epochs of the year. For a discussion of these points and further references, see [22, pp. 9-10 and 266-268] and [4, Sections 3.15 and 11.6]. Newton mentions the equation of time in the Scholium at the end of his definitions:

> Absolute time, in astronomy, is distinguished from relative, by the equation or correction of the apparent time. For the natural days are truly unequal, though they are commonly considered as equal, and used for a measure of time; astronomers correct this inequality that they may measure the celestial motions by a more accurate time. It may be, that there is no such thing as an equable motion, whereby time may be accurately measured. All motions may be accelerated and retarded, but the flowing of absolute time is not liable to any change. The duration of perseverance of the existence of things remains the same, whether the motions are swift or slow, or none at all: and therefore this duration ought to be distinguished from what are only sensible measures thereof; and from which we

deduce it, by means of the astronomical equation. The necessity of this equation, for determining the times of a phenomenon, is evinced as well from the experiments of the pendulum clock, as by eclipses of the satellites of Jupiter.

In Proposition XVII, Theorem XV, Book III of the *Principia* he says: "That the diurnal motions of the planets are uniform, and that the libration of the moon arises from its diurnal motion. The Proposition is proved from the first Law of Motion, and Cor. XXII, Prop. LXVI, Book I. Jupiter, with respect to the fixed stars, revolves in 9^h 56^m; Mars in 24^h 39^m; Venus in about 23^h; the earth in 23^h 56^m; the sun in 25 $1/2^d$, and the moon in 27^d 7^h 43^m. These things appear by the Phenomena. (...)" In Section [35] of his *System of the World* he says: "*The planets rotate around their own axes uniformly with respect to the stars; these motions are well adapted for the measurement of time.*" The importance of this statement is that Newton is presenting an operational way of measuring absolute time. The curious fact is that this measurement has to do with the diurnal rotation of the planets relative to the fixed stars, while in his definition of absolute time there should be no relation with anything external. For a general discussion of the time concept in physics see [23], [24] and [25].

Nowadays we have two other kinematical rotations of the earth. The first is the rotation of the earth relative to the background of distant galaxies. The reality of external galaxies was established by Hubble in 1924 when he determined that the nebulae are stellar systems outside the Milky Way (after finding Cepheid variables in the nebulae). We can then determine kinematically our translational and rotational velocities relative to the isotropic frame of the galaxies. This is the frame relative to which the galaxies have no translational or rotational velocity as a whole, in which the galaxies are essentially at rest relative to one another and to this frame, apart from peculiar velocities. Our angular rotational velocity relative to this frame is essentially the same as that relative to the fixed stars.

The second modern kinematical rotation of the earth is its rotation relative to the cosmic background radiation (CBR) discovered by Penzias and Wilson in 1965, [26]. This radiation has a blackbody spectrum with a characteristic temperature of 2.7 K. Although it is highly isotropic, there is a dipole anisotropy due to our velocity relative to this radiation. This motion generates Doppler shifts which are detected and measured. In this way we can, at least in principle, determine our translational and rotational velocities relative to the frame in which this radiation is isotropic.

We have indicated four different kinematical rotations of the earth. They have to do with a relative motion between the earth and external bodies or external radiation. We cannot determine by any of these means which body is really rotating, the earth or the external ones. Up to now we can adopt any point of view without problems, namely: the earth is at rest (relative to Newton's absolute space, for instance) and these bodies rotate around the earth, or these bodies are at rest and the earth spins around its axis (relative to Newton's absolute space, for instance).

In the next sections we will see how to distinguish these two points of view dynamically.

3.3.2 The Figure of the Earth

The simplest way to know that the earth is a non-inertial frame of reference is to observe its ellipsoidal form: the earth is flattened at the poles. Newton discussed this in Props. XVIII and XIX of Book III of the *Principia*:

Proposition XVIII. Theorem XVI

That the axes of the planets are less than the diameters drawn perpendicular to the axes.

The equal gravitation of the parts on all sides would give a spherical figure to the planets, if it was not for their diurnal revolution in a circle. By that circular motion it comes to pass that the parts receding from the axis endeavor to ascend about the equator; and therefore if the matter is in a fluid state, by its ascent towards the equator it will enlarge the diameters there, and by its descent towards the poles it will shorten the axis. So the diameter of Jupiter (by the concurring observations of astronomers) is found shorter between pole and pole than from east to west. And, by the same argument, if our earth was not higher about the equator than at the poles, the seas would subside about the poles, and, rising towards the equator, would lay all things there under water.

Proposition XIX. Problem III

To find the proportion of the axis of a planet to the diameters perpendicular thereto.

(...); and therefore the diameter of the earth at the equator is to its diameter from pole to pole as 230 to 229. (...)

This theoretical prediction of Newton (until that time there was no measurement of this quantity) is quite accurate compared with modern experimental determinations [3, pp. 427 and 664, note 41].

The reason for this flattening of the earth at the poles is explained in Newtonian mechanics due to the rotation of the earth relative to absolute space or to an inertial frame of reference. The earth and all frames of reference which are at rest relative to it are non-inertial. For this reason we need to introduce in the earth's frame a centrifugal force $-\omega_d^2\rho\hat{\rho}$ in order to apply Newton's laws of motion and get correct results. Here ω_d is the dynamical rotation of the earth relative to absolute space or to any inertial frame of reference. In principle it has no relation to ω_k discussed previously. In the earth's frame of reference it is this centrifugal force responsible for the flattening of the earth. In an inertial frame of reference the flattening of the earth is explained by its dynamical rotation relative to this inertial frame of reference. According to Newtonian mechanics even if the stars and distant galaxies disappeared or did not exist, the earth would still be flattened at the poles due to its rotation relative to absolute space. As we will see, relational mechanics will give a different prediction in this case.

We present here some quantitative results for this case. We will assume the earth to be composed only of water with a constant density α at any point in its interior. We will assume that the earth spins with a constant angular velocity $\vec{\omega}_d = \omega_d\hat{z}$ relative to an inertial frame of reference, where we have chosen the z axis along the axis of rotation to simplify the analysis. The equation of motion in this inertial frame for an element of mass dm of the water occupying an infinitesimal volume dV $(dm = \alpha dV)$ is given by

$$dm\vec{g} - (\nabla p)dV = dm\vec{a} \ . \tag{3.4}$$

In this equation $\vec{g}$ is the gravitational field at the point where dm is located and $-(\nabla p)dV$ is the force on dm due to the gradient of pressure p. These are the only forces acting on dm. We will solve this equation utilizing spherical coordinates $(r,\ \theta,\ \varphi)$ with an origin at the center of the earth. As the only motion of dm is a circular orbit around the z axis, its acceleration is only centripetal, given by $\vec{a} = \vec{\omega}_d \times (\vec{\omega}_d \times \vec{r}) = -\omega_d^2\rho\hat{\rho} = -\omega_d^2\rho(\hat{r}\sin\theta + \hat{\theta}\cos\theta)$, where ρ is the distance of dm to the axis z (and not to the origin, as this distance is given by r, being the relation between the two $\rho = r\sin\theta$). Moreover, $\hat{r}$, $\hat{\theta}$ and $\hat{\rho}$ are the unit vectors along the directions r and θ (spherical coordinates), and ρ (cylindrical coordinate), respectively. The pressure gradient can be expressed in cylindrical or spherical coordinates, without difficulty. To solve this equation we need an expression for the gravitational potential. As a first approximation we utilize the gravitational field of a spherically symmetrical, homogeneous distribution of matter: a total mass $M = 4\pi R^3\alpha/3$ distributed uniformly over a sphere of radius R (later on we will improve on this approximation). Utilizing Eqs. (1.6) or (1.7) it is easy to show that the gravitational field at a point $\vec{r}$ in the interior of this sphere is given by $\vec{g} = -4\pi\alpha Gr/3 = -GMr\hat{r}/R^3$. Solving the equation above, as was done in the case of a spinning bucket, yields the pressure anywhere in the fluid as given by

$$p = -\frac{GM\alpha r^2}{2R^3} + \frac{\alpha \omega_d^2 r^2 \sin^2 \theta}{2} + \frac{GM\alpha}{2R} + P_o \,, \tag{3.5}$$

where P_o is the atmospheric pressure at the North pole ($r = R$ and $\theta = 0$). The surfaces of constant pressure are ellipsoids of revolution. Taking $p = P_o$ at the equator ($\theta = \pi/2$) yields the largest distance of the water to the origin ($R_>$). That is (supposing $\omega_d^2 R^3/2GM \ll 1$ as is the case for the diurnal rotation of the earth):

$$\frac{R_>}{R} \approx 1 + \frac{\omega_d^2 R^3}{2GM} \approx 1.0017 \,. \tag{3.6}$$

To arrive at this number we put $\omega_d = 7.3 \times 10^{-5}$ s^{-1} (one day period), $R = 6.36 \times 10^6$ m, $G = 6.67 \times 10^{-11}$ Nm^2/kg^2 and $M = 6 \times 10^{24}$ kg.

This value is approximately half of what is observed by performing measurements over the earth. The problem with this calculation is that due to its rotation, a fluid earth modifies its form. In other words, it does not remain spherical but becomes approximately ellipsoidal. Therefore, the gravitational field inside and outside the rotating earth is that given by an ellipsoid. This field can be obtained utilizing the results of exercises 6.17 and 6.21 of [27]. We will not present here all calculations but only the results we obtained following this procedure.

Consider then an ellipsoid centered on the origin of the coordinate system with semi-axes a, b and c along the axes x, y and z of the coordinate system, respectively, such that $a = b = R_>$ and $c = R_< = R_>(1 - \eta)$, with $\eta \ll 1$. We will suppose once more a constant density of matter α at all points of the ellipsoid. If M is the total mass of the ellipsoid and R its average radius we have: $M = 4\pi R^3 \alpha/3 = 4\pi R_>^2 R_< \alpha/3$.

The gravitational potential energy U between two point masses m_1 and m_2 separated by a distance r is given by $U = -Gm_1 m_2/r = m_i \Phi(\vec{r}_1)$, where $\Phi(\vec{r}_1)$ is the gravitational field at the point where m_1 is located, $\vec{r}_1$, due to m_2 located at $\vec{r}_2$. Analogously, we can calculate the gravitational potential at any point in space due to the mass of the ellipsoid. The gravitational potential Φ which we found inside the ellipsoid is given by (up to first order in η):

$$\Phi = -\frac{GM}{2R^3}(3R^2 - r^2) - \frac{GM}{R^3}\frac{\eta r^2}{5}(1 - 3\cos^2 \theta) \,. \tag{3.7}$$

The potential outside the ellipsoid is given by (once more up to first order in η):

$$\Phi = -\frac{GM}{r}\left[1 + \frac{\eta R^2}{5r^2}(1 - 3\cos^2\theta)\right] . \tag{3.8}$$

The potential energy dU of a mass element dm interacting with this ellipsoid is given by $dU = dm\Phi$. The force exerted by the ellipsoid on dm is given by $d\vec{F} = -\nabla(dU) = -dm\nabla\Phi = dm\vec{g}$. Applying this in the results above yields the gravitational field inside the ellipsoid as given by (once more up to first order in η):

$$\vec{g} = -\frac{GMr}{R^3}\left\{\left[1 - \frac{2}{5}\eta(1 - 3\cos^2\theta)\right]\hat{r} - \frac{6}{5}\eta\sin\theta\cos\theta\hat{\theta}\right\} . \tag{3.9}$$

From this equation we see that the force inside the ellipsoid, for each fixed θ, grows linearly with the distance. This fact was known by Newton (*Principia*, Book I, Prop. 91, Prob. 45, Cor. III).

Outside the ellipsoid we have:

$$\vec{g} = -\frac{GM}{r^2}\left\{\left[1 + \frac{3}{5}\frac{R^2}{r^2}\eta(1 - 3\cos^2\theta)\right]\hat{r} - \frac{6}{5}\frac{R^2}{r^2}\eta\sin\theta\cos\theta\hat{\theta}\right\} . \tag{3.10}$$

The gravitational field at the surface of the ellipsoid is given by

$$\vec{g} = -\frac{GM}{R^2}\left(1 + \frac{3}{5}\eta + \eta\frac{\cos^2\theta}{5}\right) . \tag{3.11}$$

From this relation we find that the force on a point on the surface of the ellipsoid at the pole ($r = R_<$, $\theta = 0$) to the force on a point at the surface of the ellipsoid at the equator ($r = R_>$, $\theta = \pi/2$) is given by

$$\frac{F_{pole}}{F_{equator}} \approx 1 + \frac{\eta}{5} . \tag{3.12}$$

Up to now we have supposed the ellipsoid to be at rest relative to an inertial frame of reference.

At this point we will return to the problem of a spinning earth. We can then apply Eq. (3.9) to Eq. (3.4). In this case η still needs to be determined. But from the analysis of the previous case of an spinning spherical shell, we expect η to be of the order of $\omega_d^2 R^3/GM$. With (3.9) in (3.4) we obtain the following expression of the pressure p at any point inside the fluid ellipsoidal spinning earth:

$$p = -\frac{G\alpha Mr^2}{2R^3}\left(1 + \frac{4}{5}\eta\right) + \left(\frac{\omega_d^2}{2} + \frac{3}{5}\eta\frac{GM}{R^3}\right)\alpha r^2\sin^2\theta + C , \tag{3.13}$$

where C is a constant.

Equating the pressure at $r = R_<$, $\theta = 0$ with the pressure at $r = R_>$, $\theta = \pi/2$, utilizing $\eta \ll 1$, $\omega_d^2 R^3/GM \ll 1$ and the fact that η is of the same order of magnitude as $\omega_d^2 R^3/GM$ yields $\eta = 5\omega_d^2 R^3/4GM$ and:

$$\frac{R_>}{R_<} \approx 1 + \eta \approx 1 + \frac{5\omega_d^2 R^3}{4GM} \approx 1.0043 \ . \tag{3.14}$$

This is essentially the value given by Newton, $R_>/R_< \approx 230/229 \approx 1.0044$.

There are two important things to observe here. The first is that to obtain this result we utilized together the rotation of the earth and the gravitational field of an ellipsoid (the previous result (3.6) did not yield a precise value, since we assumed the gravitational field of a sphere). The second point is that the ω_d which appears in Eq. (3.14) is the dynamical rotation of the earth relative to absolute space or to an inertial frame of reference. In principle this ω_d has nothing to do with the kinematical rotation of the earth relative to the fixed stars discussed above, ω_k. But to arrive at the correct value for the flattening of the earth as observed by the measurements ($R_>/R_< \approx 1.004$) it is necessary to have $\omega_d \approx 7.3 \times 10^{-5}$ s^{-1}. In other words, ω_d needs to be equal to ω_k, or the dynamical rotation of the earth needs to have the same value as its kinematical rotation relative to the fixed stars! This should not be a coincidence; the problem is to find the connection between these two facts.

3.3.3 Foucault's Pendulum

The most striking demonstration of the rotation of the earth was obtained in 1851 by Foucault (1819-1868). The original French paper can be found in [28], while the English translation can be found in [29]. The importance of this experiment is that it can be performed in a closed room, so that we obtain the rotation of the earth without looking at the sky.

It is simply a long pendulum which oscillates to and fro many times with a long period. The pendulum is not charged and the only forces acting on it are the gravitational attraction of the earth and the tension in the string. Foucault initially utilized a pendulum with a length of 2 meters and a sphere of 5 kg oscillating harmonically. Later on he utilized another pendulum with a suspension cord of 11 meters. Although he does not mention it in the paper, soon afterward he performed his experiment at the dome of the Pantheon, with a cord of 65 meters ([29, see the footnote on page 352 by E. Fr. Jr.]). The period of an oscillating simple pendulum of length ℓ is given by $T = 2\pi\sqrt{\ell/g}$, where $g \approx 9.8$ m/s^2. Suppose the pendulum initially at rest relative to the earth, and released from an initial angle θ_o. Neglecting the effects of wind we might expect the pendulum to always oscillate in the same plane formed by the vertical direction of the weight and the direction of tension along the string. But this is not what happens. The plane of oscillation changes slowly with time relative to the earth's surface, with an angular velocity Ω. In Newtonian mechanics this is explained by means of another fictitious force, the Coriolis force given by $-2m_i\vec{\omega}_d \times \vec{v}$, where $\vec{\omega}_d$ is the angular rotation of the earth relative to an inertial system of reference (the centrifugal force does not change the plane of oscillation, so that we do not consider it here to simplify the analysis). Coriolis (1792-1843) discovered this force while doing his doctoral work under Poisson, as related in [30].

The simplest way to understand this behaviour is to consider a pendulum oscillating at the North pole. The pendulum will keep its plane of oscillation fixed relative to an inertial frame of reference (or relative to space, as it is usually termed). As the earth turns beneath it, the plane of oscillation relative to the earth changes with an angular velocity $\vec{\Omega} = -\vec{\omega}_d = -\omega_d\hat{z}$, because the earth is rotating relative to the inertial system with an angular velocity $\vec{\omega}_d = \omega_d\hat{z}$. At the equator, Foucault's pendulum does not precess because here $\vec{\omega}_d \times \vec{v}$ is either zero (when $\vec{v} = \pm v\hat{z}$) or points vertically along the length of the string (when there is a velocity component perpendicular to $\hat{z}$ and $\hat{r}$). In general the precession of the pendulum relative to the earth is given by $\Omega = -\omega_d\cos\theta$, where θ is the angle between the radial direction $\hat{r}$ (the direction in which the pendulum hangs at rest without oscillation) and the earth's axis of rotation $\vec{\omega}_d/\omega_d = \hat{z}$, shown in Figure 3.10.

We derive this result here utilizing some approximations which are valid for the problem. Accordingly, we neglect air resistance and the centrifugal force. The equation of motion in the earth's frame of reference is then given by:

$$\vec{T} + m_g\vec{g} - 2m_i\vec{\omega}_d \times \vec{v} = m_i\vec{a} \ .$$

Here $\vec{T}$ is the tension in the string. The novelty compared with the equation of motion of a simple pendulum in an inertial frame of reference is the introduction of the Coriolis force $-2m_i\vec{\omega}_d \times \vec{v}$, where $\vec{\omega}_d$ is the dynamical angular rotation of the earth relative to absolute space or to an inertial frame of reference.

We choose a new coordinate system (x', y', z') with its origin O' directly below the point of support, at the point of equilibrium of the pendulum bob, with the z'-axis pointing vertically upwards: $\hat{z}' = \hat{r}$. The x'-axis is chosen such that the pendulum would oscillate completely in the $x'z'$ plane if it were not the Coriolis force, shown in Figure 3.11.

In this frame of reference we have $\vec{\omega}_d = \omega_d\sin\theta\hat{x}' + \omega_d\cos\theta\hat{z}'$. The angle of oscillation of the pendulum with the vertical from the point of support is called β. For $\beta \ll \pi/2$ we can utilize the approximation of small amplitude of oscillation so that the equation of motion yields the approximate solution (not taking into account for the moment the Coriolis force): $\beta = \beta_o\cos\omega_o t$, where $\omega_o = \sqrt{g/\ell}$ is the natural frequency of oscillation of the pendulum and β_o

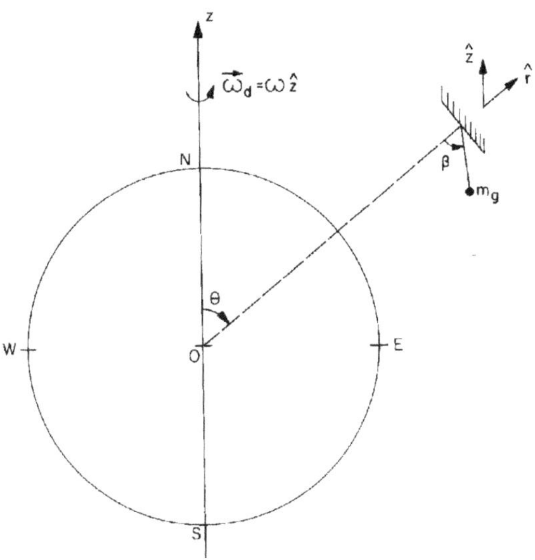

Figure 3.10: Foucault's pendulum.

the angle of release of the pendulum from rest. As we have small amplitudes of oscillation, the motion of the pendulum is essentially horizontal with $x' = \ell\beta$, so that $\vec{v} \approx -\dot{x}'\hat{x}' = \ell\beta_o\omega_o \sin\omega_o t\,\hat{x}'$. The only force component in the y' direction is given by the Coriolis force $-2m_i\vec{\omega}_d \times \vec{v}$. With the previous values for $\vec{\omega}_d$ and $\vec{v}$ we find that the equation of motion in the y' direction takes the form:

$$\ddot{y}' = -2(\omega_d \cos\theta)\ell\beta_o\omega_o \sin\omega_o t \ .$$

Integrating this equation twice and utiling the fact that $\dot{y}'(t=0) = 0$ and $y'(t=0) = 0$ yields:

$$y' = 2\omega_d \cos\theta\ell\beta_o \left(\frac{\sin\omega_o t}{\omega_o} - t \right) \ .$$

Between half a period ($t = 0$ and $t = \pi/\omega_o$) the bob moved in the y' direction an amount of $\triangle y' = -2\omega_d \cos\theta\ell\beta_o\pi/\omega_o$. During this time the bob moved in the x' direction an amount of $\triangle x' = 2\ell\beta_o$, Figure 3.12. This means that the plane of oscillation of the pendulum moved by an angle of $\triangle y/\triangle x = -\omega_d \cos\theta\pi/\omega_o$. The angular rotation Ω of the plane of oscillation is this amount divided by the time interval of $\triangle t = \pi/\omega_o - 0 = \pi/\omega_o$, so that:

$$\Omega = -\omega_d \cos\theta \ . \tag{3.15}$$

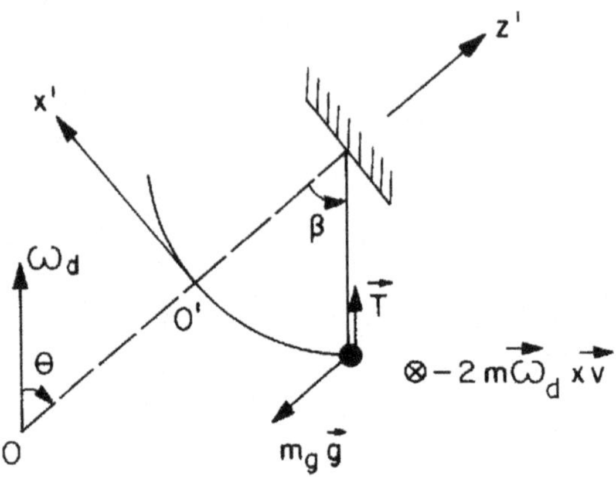

Figure 3.11: Forces in Foucault's pendulum.

Foucault did not present the calculations, but arrived at this result, stating that the angular rotation of the plane of oscillation is equal to the angular rotation of the earth multiplied by the sine of the latitude, [28] and [29]. At Paris, where Foucault performed his experiments, we have a latitude given by $\alpha = 48°51'$. As the angle of latitude is given by $\alpha = \pi/2 - \theta$, we get Foucault's result:

$$\Omega = -\omega_d \cos(\pi/2 - \alpha) = \omega_d \sin\alpha \ . \tag{3.16}$$

It is curious to note Foucault's description of his experiment. Sometimes he speaks of the rotation of the earth relative to space and other times relative to the fixed stars (heavenly sphere). He does not distinguish these two rotations or these two concepts (dynamical rotation relative to absolute space and kinematical rotation relative to the celestial bodies). For instance, he begins by stating that his experiment showing the rotation of the plane of oscillation "gives a sensible proof of the diurnal motion of the terrestrial globe." To justify this interpretation of the experimental result he imagines a pendulum placed exactly at the North pole oscillating to and fro in a fixed plane, while the earth rotates below the pendulum. He then says (our emphasis), [29]:

Thus a movement of oscillation is excited in an arc of a circle whose plane is clearly determined, to which the inertia of the mass gives an invariable position *in space*. If then these oscillations continue for a certain time, the motion of the earth, which does not cease turning from west to east, will become sensible by contrast with the

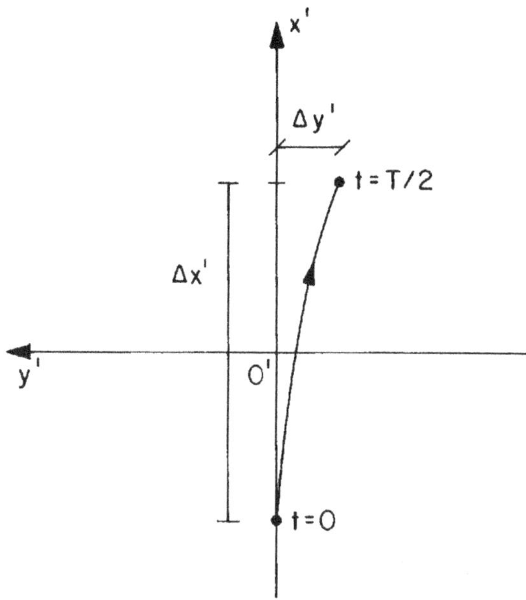

Figure 3.12: Rotation of the plane of oscillation of Foucault's pendulum.

immobility of the plane of oscillation, whose trace upon the ground will appear to have a motion conformable to the *apparent motion of the heavenly spheres*; and if the oscillations could be continued for twenty-four hours, the trace of their plane would have executed in that time a complete revolution about the vertical projection of the point of suspension.

When describing his real experiments, he states: "In less than a minute, the exact coincidence of the two points ceases to be reproduced, the oscillating point being displaced constantly towards the left of the observer; which indicates that the direction of the plane of oscillation takes place in the same direction as the horizontal component of the apparent motion of the celestial sphere."

It must be stressed that the ω_d which appears in Eqs. (3.15) and (3.16) is the dynamical rotation of the earth relative to absolute space or to any inertial frame of reference. Experimentally it is found that this ω_d has the same value (in direction and order of magnitude) as the kinematical rotation of the earth relative to the fixed stars, $\omega_d = \omega_k$. But there is no explanation of this fact in Newtonian mechanics.

The mathematical analysis leading to Eq. (2.19), $\Omega = -qB/2m_i$, was analogous to the mathematical derivation leading to Eq. (3.15), $\Omega = -\omega_d \cos\theta$. The difference is that in the first case we were in an inertial frame of reference and the precession of the plane of oscillation of the charged pendulum was due to its interaction with the magnet. On the other hand, in Foucault's pendulum we have an electrically neutral pendulum and we do not find the material agent (analogous to the magnet in the first case) responsible for the precession of the plane of oscillation. The Coriolis force $-2m_i\vec{\omega}_d \times \vec{v}$ is called a "fictitious" force because it only appears in non-inertial frames of reference which are rotating relative to absolute space. Conversely, the magnetic force $q\vec{v} \times \vec{B}$ is due to a real interaction between the charge q and the source of $\vec{B}$ (a magnet, a solenoid, a spinning charged spherical shell, *etc.*) In the earth's frame of reference we see the set of stars and galaxies rotating around us with a period of one day relative to the North-South direction (around the North pole star). In Newtonian mechanics this set of spherical shells (stars and galaxies) rotating around the earth does not generate any net force on the pendulum, whether it is at rest or moving relative to the earth. It might be thought that this set of spherical shells composed of stars and galaxies would, when rotating around the earth, generate some kind of "gravitational magnetic" field $\vec{B}_g$ which might explain the Coriolis force by a gravitational interaction analogous to the magnetic force. In other words, by an expression like $m_g\vec{v} \times \vec{B}_g$. However, even if this is the case, it cannot be due to Newton's law of universal gravitation. As we have seen, a spherical shell does not exert any force inside itself, whether the shell is at rest or rotating, no matter what the postion, velocity and acceleration of the internal test body. We will see that there is something analogous to $m_g\vec{v} \times \vec{B}_g$ in Einstein's general theory of relativity, but that it does not have exactly the same value as the Coriolis force. On the other hand, in relational mechanics this term will appear (due to the rotation of the distant matter) with the precise value of the Coriolis force. This will allow us to show that the Coriolis force is a real force due to an interaction between the test body and the rotating universe around it, contrary to what happens in Newtonian mechanics.

Max Born discussed several examples of bodies in rotation and the dynamical effects which appear. He presented the fundamental conclusion of Newtonian mechanics in simple and clear terms [31, p. 84]. In particular he emphasized that the centrifugal force of classical mechanics is universal and cannot be due to interactions between bodies, since it is due to rotation relative to absolute space.

3.3.4 Comparison of the Kinematical and Dynamical Rotations

Here we analyse these two rotations of the earth. The kinematical rotation is a relative rotation between the earth and surrounding bodies, such as the sun, the fixed stars, the distant galaxies or the CBR. The period of rotation is essentially one day ($\omega_k \approx 7 \times 10^{-5}$ rad/s) and the direction is north-south (pointing towards the north pole star in the northern hemisphere). This kinematical rotation may be equally well attributed in classical mechanics to two opposite causes: the rotation of the external world while the earth remains at rest; or a spinning of the earth around its axis while the external world does not rotate. Kinematically, we cannot distinguish between these two situations.

A completely different rotation of the earth is obtained by its flattened figure or by Foucault's pendulum. The rotation obtained dynamically by these means is a rotation of the earth relative to an inertial frame of reference. According to Newtonian mechanics, these dynamical effects (deformation of the spherical form of the earth or rotation of the plane of oscillation of the pendulum) can only be explained by a rotation of the earth relative to absolute space or to an inertial frame of reference. These effects would not appear if the earth were at rest relative to absolute space or to an inertial frame of reference, while the surrounding bodies (fixed stars and distant galaxies) were rotating in the opposite direction relative to this inertial frame. The kinematical rotation would be the same in this case, but the dynamical effects would not appear. As we will see, Mach had a different point of view, namely, that if the kinematical situation is the same, the dynamical effects must also be the same. Relational mechanics implements this quantitatively.

In classical mechanics it is a great coincidence that these two rotations happen to be the same. In other words, the rotation determined by looking at the stars kinematically is the same as the dynamical rotation determined in a closed room by Foucault's pendulum. There is no explanation for this remarkable fact in Newtonian mechanics. In the same way, there is no explanation for the equality $m_i = m_g$. Classically, we can only say that nature happens to behave this way, but a closer understanding is not supplied. The inertial mass of a body did not need to be connected to its gravitational mass. It could have been a completely independent property of the body without any relation to m_g, or it could depend on a chemical or nuclear property of the body without being in conflict with any law of classical mechanics. It only happens that experimentally the inertia is found to be proportional to the weight. A similar situation happens with the equality between the kinematical and dynamical rotations of the earth. The fact that the kinematical and dynamical rotations of the earth are the same indicates that the universe as a whole does not rotate relative to absolute space or relative to any inertial system of reference. The earth spins around its axis with a period T of one day ($T = 8.640 \times 10^4$ s), or with an angular frequency of $\omega = 2\pi/T = 7 \times 10^{-5}$ rad/s. The earth orbits around the sun with a period of one year ($T = 3.156 \times 10^7$ s), or an angular frequency of $\omega = 2 \times 10^{-7}$ rad/s. The planetary system orbits around our galaxy with a period of 2.5×10^8 years, ($T = 8 \times 10^{15}$ s), or an angular frequency of

$\omega \approx 8 \times 10^{-16}$ rad/s. Most astronomical bodies in the universe rotate, except the universe as a whole. Why does the universe as a whole not rotate relative to absolute space? There is no explanation for this fact in classical mechanics. This is a fact of observation, but nothing in classical mechanics obliges nature to behave like this. The laws of mechanics would remain the same if the universe as a whole were rotating relative to absolute space. We would only need to take this into account when performing calculations (this would cause a flattening in the distribution of galaxies, similar to the essentially plane form of the solar system or of our galaxy due to their rotation).

These two coincidences of classical mechanics ($m_i = m_g$ and $\vec{\omega}_k = \vec{\omega}_d$) form the main empirical foundations for Mach's principle.

3.4   General fictitious Force

In an inertial frame S we can write Newton's second law of motion as

$$m \frac{d^2 \vec{r}}{dt^2} = \vec{F} \; ,$$

where $\vec{r}$ is the position vector of the particle m relative to the center O of S.

Suppose now we have a non-inertial frame of reference S' which is located by a vector $\vec{h}$ with respect to S ($\vec{r} = \vec{r}' + \vec{h}$, where $\vec{r}'$ is the position vector of m relative to the origin O' of S'), moving relative to it with translational velocity $d\vec{h}/dt$ and translational acceleration $d^2\vec{h}/dt^2$. Suppose, moreover, that the axes x', y', z' rotate relative to the axes x, y, z of S with an angular velocity $\vec{\omega}$. In this frame S' taking into account the complete "fictitious forces" Newton's second law of motion should be written as ([27], Chapter 7):

$$m \frac{d^2 \vec{r}\,'}{dt^2} = \vec{F} - m\vec{\omega} \times (\vec{\omega} \times \vec{r}\,')$$

$$- \, 2m\vec{\omega} \times \frac{d\vec{r}\,'}{dt} - m \frac{d\vec{\omega}}{dt} \times \vec{r}\,' - m \frac{d^2 \vec{h}}{dt^2} \; . \tag{3.17}$$

The second term on the right is called the centrifugal force, the third term is called the Coriolis force, the fourth and fifth terms have no special names. They are all "fictitious forces" in Newtonian mechanics, and appear only in non-inertial frames of reference. Although their effects are real in these non-inertial frames (flattening of the earth, concave form of the water in Newton's bucket experiment, Foucault's pendulum, ...), we cannot find a physical origin for these forces. That is, we cannot find the body responsible for them and the possible nature of this interaction (if gravitational, electric, magnetic, elastic, nuclear, *etc.*) Certainly in classical mechanics these fictitious forces are not caused by the fixed stars or by the external galaxies. The reason is that even if the stars or galaxies disappeared or doubled in number and mass, the fictitious forces would still be there with the same values in any non-inertial frame of reference. Hence the name "fictitious" forces.

Chapter 4

Gravitational Paradox

In this chapter we discuss the gravitational paradox. References can be found at: [32], Chapter 2 (Cosmological Difficulties with the Newtonian Theory of Gravitation), pp. 16-23; [33], pp. 194-195; [34], Chapter 8 (The Gravitational Paradox of an Infinite Universe), pp. 189-212; [35]; [36]; [12], Chapter 7, Sections 203-222.

4.1 Newton and the Infinite Universe

The cosmological conceptions of Isaac Newton have been clearly analysed by E. Harrison in an interesting paper [37]. Harrison's work shows that during his early years (1660's), Newton believed that space extended infinitely in all directions and was eternal in duration. The material world, on the other hand, was of a finite extent. It occupied a finite volume of space and was surrounded by an infinite space devoid of matter.

After his complete formulation of universal gravitation, Newton became aware that the fixed stars might attract one another due to their gravitational interaction. In the General Scholium at the end of the *Principia* Newton wrote: "This most beautiful system of the sun, planets, and comets, could only proceed from the counsel and dominion of an intelligent and powerful Being. And if the fixed stars are the centres of other like systems, these, being formed by the like wise counsel, must be all subject to the dominion of One; especially since the light of the fixed stars is of the same nature with the light of the sun, and from every system light passes into all the other systems: and lest the systems of the fixed stars should, by their gravity, fall on each other, he hath placed those systems at immense distances from one another." However, putting the fixed stars very far away from one another does not avoid another problem: if the universe existed for an infinite amount of time, then a finite amount of matter occupying a finite volume would eventually collapse to its center due to the gravitational attraction of the inner matter.

In correspondence exchanged with the theologian Richard Bentley in 1692-93, Newton perceived this fact and changed his cosmological views. He abandoned the idea of a finite material universe surrounded by an infinte void, and defended the idea of an infinite material world spread out in infinite space. This can be seen in his first letter to Bentley [38, p. 281].

With an infinite amount of matter distributed more or less homogeneously over the whole of an infinite space, there would be approximately the same amount of matter in all directions. In this way there would be no center of the world to where the matter would collapse. Two hundred years later, however, a paradoxical situation was identified with this cosmological system. This is the subject of the next sections.

4.2   The Force Paradox

There is a simple but profound paradox which appears with Newton's law of gravitation in an infinite universe which contains an infinite amount of matter. The simplest way to present the paradox is the following: Suppose a boundless universe with an homogeneous distribution of matter. We represent its constant density of gravitational mass by ρ. To simplify the analysis we deal here with a continuous mass distribution extending uniformly to infinity in all directions. We now calculate the gravitational force exerted by this infinite universe on a test particle with gravitational mass m located at a point P, as in Figure 4.1.

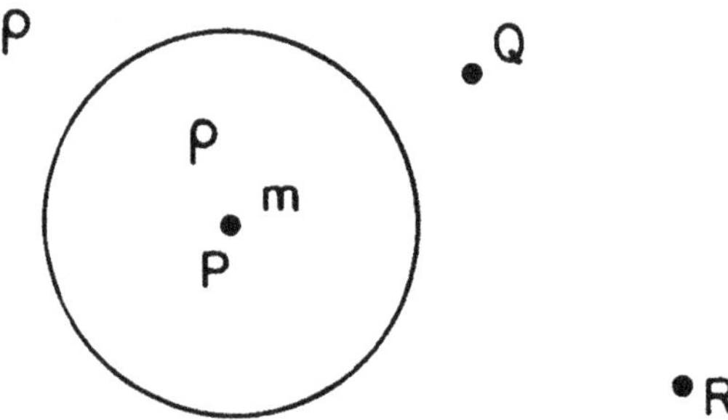

Figure 4.1: Infinite and homogeneous universe with a constant mass density ρ.

If we calculate the force with our coordinate system centered on P, all the universe will be equivalent to a series of spherical shells centered on P. From

Eq. (1.6) we learn that there will be no net force acting on m. This might be expected by symmetry.

Now let us calculate the force on m utilizing a coordinate system centered on another point Q, as in Figure 4.2.

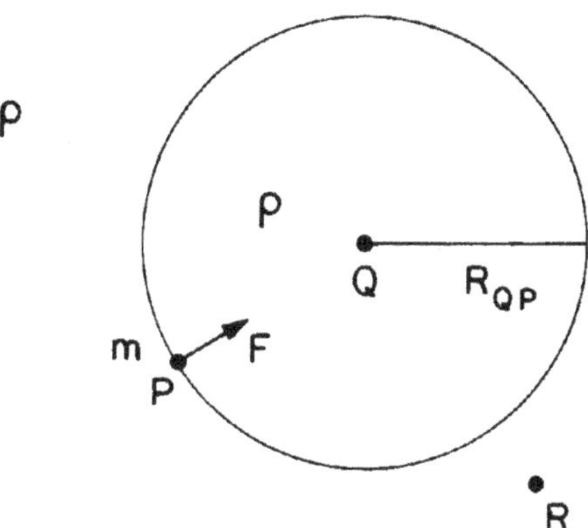

Figure 4.2: Force on m calculated from Q.

In order to calculate the net force we divide the universe into two parts centered on Q. The first one is the sphere of radius R_{QP} centered on Q and passing through P. The mass of this sphere is $M = \rho 4\pi R_{QP}^3/3$. It attracts the material point m with a force given by $GMm/R_{QP}^2 = 4\pi G\rho m R_{QP}/3$ pointing from P to Q. The second part is the remainder of the universe. This remainder is composed of a series of external shells centered on Q containing the internal test particle m. By Eq. (1.6) this second part exerts no force on m. This means that the net force exerted on m by the whole universe calculated in this way is proportional to the distance R_{QP} and points from P to Q.

Following a similar procedure but utilizing a coordinate system centered on another point R, as in Figure 4.3, we would find that the net force exerted by the whole universe on m is proportional to the distance between P and R, pointing from P to R: $F = 4\pi G\rho m R_{RP}/3$.

This means that depending on how we perform the calculation we obtain a different result. This is certainly unsatisfactory.

However, the problem is not with the mathematics. For instance, if we were calculating the force exerted by a finite distribution of mass on a test particle m utilizing Newton's law of gravitation, the result would be the same no matter how we calculated the result or where we centered the coordinate system. We assume, for instance, the finite body with constant density ρ of Figure 4.4. It is surrounded by an infinite void space. If we calculate the net gravitational

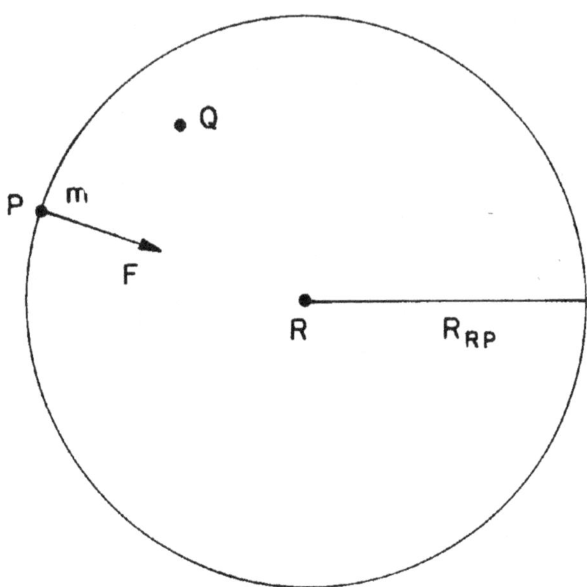

Figure 4.3: Force on m calculated from R.

force exerted by this body on one of its particles of mass m (or ρdV, where dV is the infinitesimal volume of the particle) located at T, we always obtain the same result pointing from T to S. We can perform the calculations placing the coordinate system centered on S, on T, on U, on V or on any other point, and the final result will always be the same: a force of the same magnitude pointing from T to S.

Another way of presenting the paradox is to consider the force on m_g located at P calculated from an origin at Q, shown in Figure 4.2. As we have seen, the net force on m_g points from P to Q and is proportional to the distance PQ. This means that the net force on a material particle located on P becomes infinite if it is located at an infinite distance from Q.

This is called the gravitational paradox. It was discovered by Seeliger and Carl Neumann at the end of last century (1894-96).

4.3   The Paradox based on Potential

Instead of calculating the force, we could just as well calculate the gravitational potential or the gravitational potential energy.

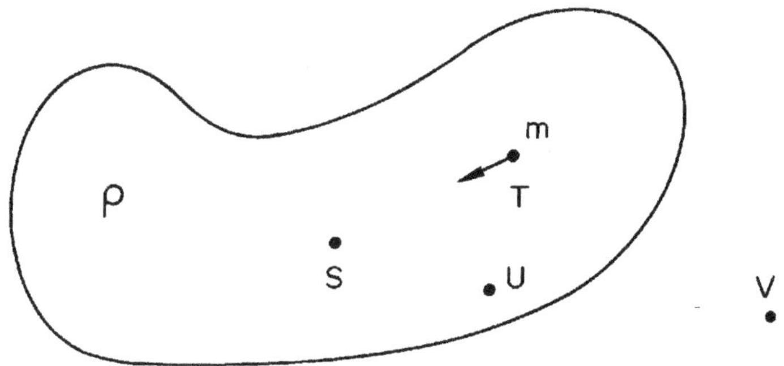

Figure 4.4: Finite body attracting one of its particles.

The gravitational potential at a point $\vec{r}_o$ due to N gravitational masses m_{gj} located at $\vec{r}_j$ is given by:

$$\Phi(\vec{r}_o) = -\sum_{j=1}^{N} G\frac{m_{gj}}{r_{oj}} \,,$$

where $r_{oj} \equiv |\vec{r}_o - \vec{r}_j|$. The gravitational energy of a material particle m_{go} located at $\vec{r}_o$ interacting with these N masses is given by $U = m_{go}\Phi$.

We now calculate the gravitational potential at a point $\vec{r}_o = r_o\hat{z}$ due to a spherical distribution of mass of radius $R > r_o$, thickness dR and mass $dM_g = 4\pi R^2 dR\rho_g$ (ρ_g being the uniform volumetric density of mass of the shell) with the previous expression. Substituting the sum by a surface integral over the shell and m_{gj} by $d^3M_g = \rho_g R^2 dRd\varphi \sin\theta d\theta$ yields the well-known result

$$d\Phi(r_o < R) = -G\rho_g R^2 dR \int_{\varphi=0}^{2\pi} \int_{\theta=0}^{\pi} \frac{\sin\theta d\theta d\varphi}{\sqrt{R^2 + r_o^2 - 2Rr_o\cos\theta}}$$

$$= -\frac{GdM_g}{R} = -4\pi G\rho_q R dR \;.$$

The contribution of the shell is proportional to the radius of the shell. This means that if we integrate this from $R = 0$ to infinity we obtain an infinite result. This was obtained by Seeliger and Neumann. The same can be said of the gravitational energy of a point particle interacting with this infinite homogeneous universe, *i.e.*, it becomes infinite. The force on the test particle should be obtained by minus the gradient of this potential energy, but this becomes indefinite.

There is another way to present the paradox. The equation satisfied by the gravitational potential in the presence of matter is known as Poisson's equation:

$$\nabla^2 \Phi = 4\pi G \rho_g \; .$$

This is easily obtained observing that $\nabla^2(1/r) = -4\pi\delta(\vec{r})$, where $\delta(\vec{r})$ is Dirac's delta function. Utilizing $\Phi = -Gm_g/r$ and the fact that $m_g\delta(\vec{r}) = \rho_g(\vec{r})$ yields this equation.

If we have a homogeneous universe with a constant density of mass we should expect a constant Φ. But supposing Φ to be a finite constant yields $\rho_g = 0$ from this equation, which is against the initial supposition of a constant and finite density different from zero. There is no solution of Poisson's equation with a constant Φ and a constant ρ_g different from zero.

4.4 Solutions of the Paradox

There are three main ways of solving the paradox: (I) The universe has a finite amount of mass. (II) Newton's law of gravitation should be modified. (III) There are two kinds of mass in the universe, positive and negative.

(I) In the first solution we maintain Newton's law of gravitation and the constituents of the universe as usually known. We only require a finite amount of matter in order to avoid the paradox. For instance, if the universe has a total finite mass M uniformly distributed around a center P of radius R, with a constant mass density $\rho = M/(4\pi R^3/3)$ the net gravitational force exerted on a test particle m located at $r < R$ is given simply by $Gm(4\rho\pi r^3/3)/r^2 = 4\pi G\rho mr/3$ pointing from m to P, no matter how we perform the calculation. We can center the coordinate system on P, on m or at any other point, and the final result will be the same.

However, this solution creates other problems. As we have seen, Newton abandoned this cosmological model of the universe because it leads to a collapsing situation. The external matter tends to concentrate on the center due to the gravitational attraction of the inner matter. To avoid this problem we would need to suppose the universe to be rotating relative to absolute space (the planetary system does not collapse into the sun due to its rotation, so that the centripetal gravitational force of the sun is balanced by $m_i\vec{a}$, or by a centrifugal force in a frame of reference rotating with the planets). But we saw previously that the universe as a whole does not rotate relative to absolute space (the best inertial frame we have is the frame in which the distant galaxies are seen without rotation). This means that this proposed solution would be refuted by observations. We would then need to postulate some kind of repulsive force as yet unknown to avoid the gravitational collapse of the finite universe.

(II) The second solution was proposed by Seeliger and C. Neumann in 1895-96. Essentially, they proposed that the gravitational potential $\Phi = -Gm/r$

should be replaced by $-Gme^{-\alpha r}/r$, where α has dimensions of length^{-1} and gives the typical range of interaction (the order of magnitude up to where gravitation is really effective). It should be stressed that Seeliger and Neumann proposed this potential 50 years before Yukawa suggested a similar law describing nuclear interactions. If we have two interacting bodies m_{g1} and m_{g2} separated by a distance r_{12} their gravitational potential energy would be given by

$$U = -G\frac{m_{g1}m_{g2}}{r_{12}}e^{-\alpha r_{12}} .$$

From this point on, we present our own calculations. Utilizing the fact that $\vec{F} = -\nabla U$ we can obtain the force exerted by m_{g2} on m_{g1}, assuming α to be a constant:

$$\vec{F} = -G\frac{m_{g1}m_{g2}}{r_{12}^2}\hat{r}_{12}(1 + \alpha r_{12})e^{-\alpha r_{12}} . \tag{4.1}$$

We now integrate this equation, assuming a universe with constant mass density ρ_2. The test particle of gravitational mass m_{g1} is located on the z-axis at a distance d_1 from the origin of the coordinate system at O, $\vec{r}_1 = d_1\hat{z}$. We consider an element of mass dm_{g2} located at $\vec{r}_2 = r_2\hat{r}_2$. Once more we divide the universe in two parts centered at O: The first part is at $r_2 > d_1$ while the second is at $r_2 < d_1$, shown in Figure 4.5.

We now integrate the gravitational force exerted by a spherical shell of radius r_2 on m_{g1} utilizing spherical coordinates, with φ_2 going from zero to 2π and θ_2 going from zero to π. With $r_2 > d_1$ we get:

$$d\vec{F} = \frac{2\pi Gm_{g1}\rho_2 r_2 e^{-\alpha r_2} dr_2\hat{z}}{d_1^2\alpha}\left[(1 + \alpha d_1)e^{-\alpha d_1} - (1 - \alpha d_1)e^{\alpha d_1}\right] .$$

This is different from zero if $d_1 \neq 0$. This means that a spherical shell will exert a net force on an internal test particle according to Seeliger-Neumann's potential if it is not at the center. There is a striking difference between this result and the Newtonian case, which yields zero no matter the position of the internal test particle.

In the limit in which $\alpha d_1 \ll 1$ we recover the Newtonian result where a spherical shell exerts no force on a test particle localized anywhere inside the shell.

Integrating this result from $r_2 = d_1$ to infinity yields:

$$\vec{F} = -\frac{2\pi Gm_{g1}\rho_2(1 + \alpha d_1)\hat{z}}{d_1^2\alpha^3}\left[(1 - \alpha d_1) - (1 + \alpha d_1)e^{-2\alpha d_1}\right] . \tag{4.2}$$

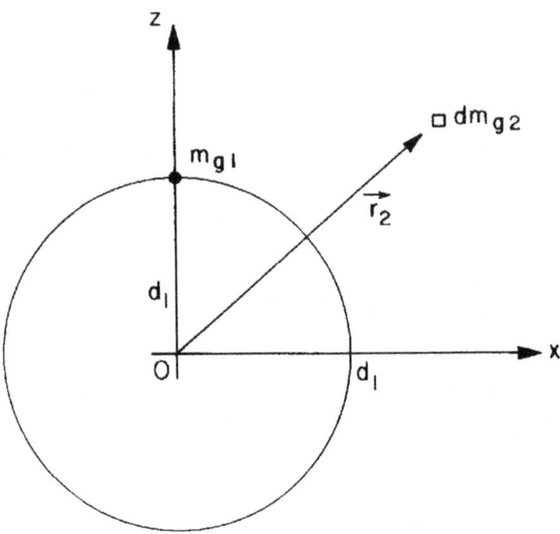

Figure 4.5: Coordinate system to calculate the force on m_{g1}.

We now calculate the force on m_{g1} due to the second part $r_2 < d_1$. We first calculate the force of a spherical shell attracting an external particle. Integrating Eq. (4.1) in φ_2 going from zero to 2π and in θ_2 going from zero to π yields, with $r_2 < d_1$:

$$d\vec{F} = \frac{2\pi G m_{g1}\rho_2(1 + \alpha d_1)e^{-\alpha d_1} r_2 dr_2 \hat{z}}{d_1^2 \alpha} \left(e^{-\alpha r_2} - e^{\alpha r_2}\right) \ .$$

In the limit in which $\alpha r_2 \ll 1$ and $\alpha d_1 \ll 1$, we recover the Newtonian result that a spherical shell attracts an external point as if it were concentrated at its center, namely:

$$d\vec{F} = -\frac{4\pi r_2^2 dr_2 \rho_2 G m_{g1}\hat{z}}{d_1^2} \ .$$

Integrating the previous result in r_2 going from zero to d_1 yields:

$$\vec{F} = \frac{2\pi G m_{g1}\rho_2(1 + \alpha d_1)\hat{z}}{d_1^2 \alpha^3} \left[(1 - \alpha d_1) - (1 + \alpha d_1)e^{-2\alpha d_1}\right] \ . \tag{4.3}$$

This result is valid for $\alpha \neq 0$ and cannot be applied for $\alpha = 0$.

Adding Eqs. (4.2) and (4.3) yields zero as the resultant force acting on m_{g1} due to the whole universe. The same result is obtained choosing any other point as the origin of the coordinate system. This shows that the paradox is solved with the Seeliger-Neumann potential energy, even keeping an infinite and homogeneous universe.

We now analyse this solution of the paradox as regards the potential. The equation satisfied by a potential $\Phi = -Gm_g e^{-\alpha r}/r$ due to a point mass m_g is given by:

$$\nabla^2 \Phi - \alpha^2 \Phi = 4\pi G \rho_g \ .$$

There is now a solution for this equation with a constant and finite ρ_g yielding a constant and finite $\Phi = -4\pi G\rho_g/\alpha^2$. With the known values of G, ρ_g and utilizing $\alpha = H_o/c$ (H_o is Hubble's constant) yields a potential close to $-c^2$.

Another way of obtaining this result is directly integrating the potential due to a spherical shell of radius r. To this end we replace m_g by $d^3 m_g = \rho \sin\theta d\theta r^2 dr d\varphi$ and calculate the potential at the origin. Integrating:

$$\Phi = -G\rho_g \int_{r=0}^{\infty} \int_{\theta=0}^{\pi} \int_{\varphi=0}^{2\pi} (r^2 \sin\theta dr d\theta d\varphi) \frac{e^{-\alpha r}}{r} = -\frac{4\pi G\rho_g}{\alpha^2} \ .$$

The same result is obtained calculating the potential at any other point $\vec{r}_o$ different from the origin.

This shows the solution of the gravitational paradox based on the potential.

(III) The third way of solving the gravitational paradox is to suppose the existence of negative gravitational masses. The first to propose this idea of a negative gravitational mass seems to have been Föppl in 1897 [39, p. 234]. He proposed this based on electromagnetic analogies, and was not concerned with the gravitational paradox. Calling the ordinary mass positive, we would have the following rule: positive mass attracts positive mass but repels negative mass, while negative mass attracts negative mass and repels positive mass. This would be the opposite of what happens with electrical charges. This being the case, we could have a universe with an equal amount of positive and negative masses, in which Newton's law of gravitation would be obeyed and in which the gravitational paradox would not appear, even with an infinite amount of positive mass. Now there is a solution of the equations in which both masses are equally distributed everywhere, so that the net gravitational force on any body is zero on the average. The gravitational potential energy would also be zero everywhere on the average. There exists a solution of Poisson's equation with a constant Φ and a zero ρ_g.

We can understand this third solution more easily observing that there is no electrical paradox analogous to the gravitational paradox. The reason is that usually we consider the universe as a whole to be electrically neutral. In other words, apart from local anisotropies, the negative charge in one region is compensated by a corresponding positive charge somewhere else. This means that on average there is no electrostatic force on any charge due to all the charges in the universe. The same would be true for gravitation, provided there is negative gravitational mass.

The gravitational paradox is very simple to state and understand. It is amazing that with a situation so simple we can arrive at such a far-reaching conclusion, namely: We cannot have a universe with an infinite amount of ordinary matter in which Newton's law of gravitation is obeyed. At least one of these components must be modified: The infinite amount of matter in the universe, Newton's law of gravitation, or the nature of the constituents in the cosmos (if we have negative masses).

Our own preferred cosmological model is a universe that is boundless and infinite in space, which has always existed without any creation, and with an infinite amount of matter in all directions. In this model the universe extends in all directions without end, with an infinite amount of matter on the whole, but with a finite matter density on average. The simplest universe model along these lines is an homogeneous distribution of mass in the large scale with a finite matter density. This means that it has no preferred center, so that any point can be arbitrarily chosen as its center. We could also perform the calculations beginning from any point. For this reason we do not adopt the first solution of the gravitational paradox. We prefer the second and third solutions. In this book we explore quantitatively only the second solution.

4.5   Absorption of Gravity

There have been other reasons for people to propose an exponential decay in the gravitational potential of a point mass, in the gravitational potential energy between two point masses, or in the gravitational force. These ideas are not directly related to the gravitational paradox, but sometimes the proposed modification is along the same lines. We have reviewed this situation elsewhere ([35], [36] and [12, Sections 7.5 to 7.7]); all the references and further discussion can be found in these studies. In each paragraph below we discuss a different kind of idea leading to an exponential decay for gravitation.

Light flowing from a source is absorbed by an intervening medium, so that its power falls as $e^{-\lambda r}/r^2$. Those who suppose that gravitation propagates from a source like light (in the form of gravitational waves or in the form of particles like gravitons) are led to propose an exponential decay of gravitation.

When we interpose a dielectric between two point charges, this medium becomes affected by the charges and becomes electrically polarized. The effect of this polarization is to change the net force on each of the charges, as compared with the situation in which there was no medium interposed between them. In this case we don't need to speak of a propagation of the electric force, and the situation can be described by a simultaneous many-body interaction. Nevertheless, if we assume an analogy between electromagnetism and gravitation, we might expect some influence of the intervening medium for the net gravitational force on any material body. We may once more need to introduce an exponential decay for gravitation, although in this case there is nothing propagating at a finite speed. The only thing which happens here is that an action at a distance between many bodies may have this behaviour.

Astronomical observations, such as the flat rotation curves of spiral galaxies, also led people to propose modifications in Newton's law of gravitation or to postulate the existence of dark matter. Discussions of these topics can be found elsewhere ([40] and [41]). The problem of the flat rotation curves of galaxies can be understood in a simple way. Let us suppose a gravitational interaction between a large body of mass M and a small body of mass $m \ll M$ describing a circular orbit around M in an inertial frame of reference. With Newton's law of universal gravitation, his second law of motion and the expression for the centripetal acceleration we get: $GMm/r^2 = ma = mv^2/r = m\omega^2 r$. Here r is the distance between the bodies, v is the tangential velocity of m, a is its centripetal acceleration and $\omega = v/r$ is the angular velocity of m around M. From this expression we obtain $v = \sqrt{GM/r}$ and $\omega = \sqrt{GM/r^3}$. With a larger r we have a smaller velocity v. This prediction is perfectly corroborated in the case of the planetary system, with M the sun and m any one of the planets. These relations of v and ω as a function of r are another form of Kepler's third law in the case of a circular orbit (the square of the period of revolution is proportional to the cube of the radius): $T^2 = (2\pi/\omega)^2 = 4\pi^2 r^3/GM = Kr^3$. On the other hand, this relation is not valid for galaxies. Let m be a star belonging to a galaxy and far from its center, and M the mass of the nucleus of the galaxy (determined from its visible or bright part). Observations indicate that in most galaxies the tangential velocity v becomes approximately constant as r increases, instead of falling as $1/\sqrt{r}$ (as would be expected according to Newtonian mechanics). To solve this problem there are two main approaches. One is to suppose the existence of dark matter not yet observed in any wavelength, that could interact gravitationaly with the stars. From the observed flat rotation curves the distribution of this supposed dark matter can be estimated, assuming the validity of Newton's laws. Another approach is to suppose that all existing matter has already been detected and then try to find a modification in Newtonian mechanics in order to explain the flat rotation curves. Usually the modification should be relevant for distances of the order of 10^{20} m, which is the typical size of a galaxy. We can try to modify either Newton's second law of motion $m_i a$ or the gravitational force GMm/r^2. In this last case some trials have been made with relative success based on an exponential decay for gravitation [42], [43] and [44]. The main problem with this approach is how to derive simultaneously the flat rotation curves of galaxies and Tully-Fisher's law (luminosity proportional to the square of the tangential velocity of a galaxy). An alternative model has been developed elsewhere ([45] and [46]). Although it does not deal with absorption of gravity nor with its exponential decay, it leads to effects involving an exponential decay which has some mathematical analogies with what is being discussed here. Further research is necessary before we can draw a final explanation for the flat rotation curves of galaxies.

Some terrestrial experiments have been made to detect modifications in Newton's law of gravitation. Some of these have met with positive results, as those of Majorana in the beginning of this century. For this reason they should be repeated (see [47], [48], [49], [50] and [51]).

We have already discussed this problem in the references cited above, and will not analyse the subject further here. What should be stressed is that Newton's law of gravitation or any other expression may be approximately valid in some conditions, although it may be necessary to modify it due to observations of astronomical bodies or terrestrial laboratory experiments. It is important to be open-minded in this regard.

For further discussions and references on all these topics can be found elsewhere ([16] and [52]).

Chapter 5

Leibniz and Berkeley

Before we present Mach's criticisms of Newtonian mechanics we discuss the points of view of G. W. Leibniz and of the Bishop G. Berkeley as regards absolute and relative motion. These philosophers anticipated many points of view later advocated by Mach.

5.1   Leibniz and Relative Motion

Leibniz (1646-1716) was introduced to the modern sciences of his time by C. Huygens (1629-1695). They were in close contact during Leibniz's stay in Paris during 1672-1676. Huygens may have influenced him as regards the concepts of space and time, and the significance of centrifugal force. A detailed study of Huygens reactions to Newtonian mechanics can be found elsewhere ([33, pp. 119-126] and [53]).

Leibniz never accepted Newton's concepts of absolute space and time. He maintained that space and time depend on things, with space being the order of coexistent phenomena and time the order of successive phenomena. There is a very interesting correspondence (in the years 1715-1716) between Leibniz and S. Clarke (1675-1729), a disciple of Newton. Leibniz wrote in French and Clarke in English. This correspondence illuminates this whole issue and can be found in English [54].

In the fourth paragraph of his third letter to Clarke, Leibniz wrote:

> 4. As for my opinion, I have said more than once, that I hold space to be something merely relative, as time is; that I hold it to be an order of coexistences, as time is an order of successions. For space denotes, in terms of possibility, an order of things which exist at the same time, considered as existing together; without enquiring into

> their manner of existing. And when many things are seen together,
> one perceives that order of things among themselves.

Leibniz defends the idea that all motion is relative. He nevertheless admits that it may be more practical or convenient to say that some large collections of bodies which remain at rest relative to one another (like the fixed stars) are at rest, while one body moves relative to them, than to say the opposite. But this is more a matter of convention than of physical reality. This can be seen in a text written in 1689 entitled *On Copernicanism and the Relativity of Motion* [55, pp. 90-92 and 130-131].

In this text he considers that the hybrid geocentric astronomical system of Tycho Brahe may be "more appropriate for a given purpose" than the heliocentric system of Copernicus, while the Copernican system may be more appropriate for another purpose than the Tychonic one. He goes on to say that the Ptolemaic geocentric system is the truest in spherical astronomy (that is, more intelligible), while the Copernican account is most appropriate (*i.e.*, the most intelligible) for explaining the theory of the planets.

The same point of view is expressed in his work *A Specimen of Dynamics* [*Specimen Dynamicum*] of 1695 [55, p. 125].

Later on we will see that Mach also defended the idea that the Copernican and Ptolemaic systems are equally valid and correct. The only difference is that the Corpernican system is more economical or practical.

But how does Leibniz cope with Newton's key experiments of the spinning bucket and the two globes? In a letter written to Huygens in 1694 he proposed that force may be something real [55, p. 308].

At the same time that he advocates a relational theory of space and time, he seems to attach some reality or absolute value to the force or kinetic energy. This is somewhat contradictory. Nor does he state explicitly his reasons for believing that nothing breaks the general law of equivalence (the relational theory).

Here is what he says in the 53rd paragraph of his fifth letter to Clarke [54]:

> 53. I find nothing in the Eighth Definition of the *Mathematical Principles of Nature*, nor in the Scholium belonging to it, that proves, or can prove, the reality of space in itself. However, I grant there is a difference between an absolute true motion of a body, and a mere relative change of its situation with respect to another body. For when the immediate cause of the change is in the body, that body is truly in motion; and then the situation of other bodies, with respect to it, will be changed consequently, though the cause of that change be not in them. 'Tis true that, exactly speaking, there is not any one body, that is perfectly and entirely at rest; but we frame

> an abstract notion of rest, by considering the thing mathematically. Thus have I left nothing unanswered, of what has been alleged for the absolute reality of space. And I have demonstrated the falsehood of that reality, by a fundamental principle, one of the most certain both in reason and experience; against which, no exception or instance can be alleged. Upon the whole, one may judge from what has been said that I ought not to admit a moveable universe; nor any place out of the material universe.

The difficulty here has already been pointed out by H. G. Alexander: "There is, however, no doubt that this admission of the distinction between absolute and relative motion is inconsistent with his general theory of space" [54, p. xxvii]. Leibniz was led astray by Newton's arguments concerning the bucket and two globes experiments. He is here tacitly admitting that absolute motion does in fact exist, contrary to his beliefs. One way out of the contradiction would be to maintain that these effects (the concave form of the water or the tension in the string) are due to the relative rotation between the water and the globes with respect to the fixed stars. He could say that these effects appear not only when the water and globes rotate relative to the stars, but that they would also appear when the water and globes were at rest (relative to an observer or to the earth) while the stars were rotating in the opposite direction relative to them with the same angular velocity. If he had seen this possibility clearly, he could have maintained that even these experiments do not prove the existence of absolute space. He could then also maintain that the water does not need to be absolutely in motion when its surface is concave, as we might say that this would happen with the water at rest and the stars rotating around it. But Leibniz did not explicitly mention this possibility. For this reason he could not give a clear answer to the Newtonian arguments utilizing his relational theory of motion. Erlichson has already pointed this out: "In my opinion Leibniz never really answered Clarke and Newton on the bucket experiment or the other examples they give to show the dynamical effects of absolute motion" [56].

In at least one point in the correspondence, Clarke saw better than Leibniz the consequences of a completely relational theory of motion as regards the origin of the centrifugal force. In his fifth reply to Leibniz, Clarke wrote (paragraphs 26-32, p. 101 of [54]):

> It is affirmed [by Leibniz], that motion necessarily implies a (§31) relative change of situation in one body, with regard to other bodies: and yet no way is shown to avoid this absurd consequence, that then the mobility of one body depends on the existence of other bodies; and that any single body existing alone, would be incapable of motion; or that the parts of a circulating body, (suppose the sun,)

would lose the *vis centrifuga* arising from their circular motion, if
all the extrinsic matter around them were annihilated.

Unfortunately Leibniz could not respond to this last argument as it was
transmitted to him on October 29th, 1716 and he died on November 14th, 1716.
In any case, the consequences which Clarke called "absurd" are the main parts
of any real relational theory of motion. If we follow fully a relational theory of
motion, there is no sense to the notion of a single body moving relative to space;
motion is relative to other bodies. Thus, the motion of one body depends on the
existence of other bodies. Much more important is the consequence, pointed
out by Clarke, that the centrifugal force would disappear if the external bodies
were annihilated. In other words, if the stars (and galaxies) were annihilated,
the earth would not be flattened at the poles; the water would not rise towards
the sides of the bucket when they rotated; there would not appear any tension
in the string connecting the two spinning globes; it would not be possible for
the planets to orbit around the sun, as there would not exist anything to oppose
the gravitational attraction between them and the solar system would collapse
etc. This is a necessary consequence of a completely relational theory, and is
implemented in relational mechanics, as we show in this book. It is not an
"absurd" consequence of a relative theory of motion. In principle this idea
can be tested experimentally: consider a bucket with water and two globes
connected by a spring at rest relative to the earth. If we surround them by
a massive spherical shell and rotate only the shell relative to the earth, there
should appear a small centrifugal force acting on the water and on the globes
if Leibniz's ideas are correct. Later on we present orders of magnitude for this
effect based on relational mechanics. Clarke was the first to point out clearly
the fact that in a purely relational theory, the centrifugal force only appears
when there is a relative rotation between the test body and the distant material
universe. If the distant universe is annihilated, the dynamical consequences
of this centrifugal force should disappear accordingly. Other people were not
impressed by (or did not realize the significance of) these consequences pointed
out by Clarke.

Leibniz believed that kinematically equivalent motions should be dynamically equivalent. This is evident from his statement quoted above that "we must
hold that however many bodies might be in motion, one cannot infer from the
phenomena which of them really has absolute and determinate motion or rest.
Rather, one can attribute rest to any one of them one may choose, and yet the
same phenomena will result." Despite this belief, he did not implement this
idea quantitatively. For instance, he did not show how a spinning set of stars
can generate centrifugal forces. Nor did he mention the proportionality between
inertial and gravitational masses (or between inertia and weight). Finally he
did not even hint at the possibility that the centrifugal forces might have a

gravitational origin.

Although he advocated certain ideas which clashed with Newtonian mechanics, he did not develop them mathematically. The level of knowledge of physical science at that time, and especially of electromagnetism, was not yet sufficient to supply the key to implementing these ideas quantitatively.

5.2   Berkeley and Relative Motion

Berkeley (1685-1753) criticised Newton's concepts of absolute space, absolute time and absolute motion mainly in Sections 97-99 and 110-117 of his work *The Principles of Human Knowledge* (1710) and in Sections 52-65 of his work *Of Motion - Or the principle and nature of motion and the cause of the communication of motions* (1721). This work is usually known by its Latin title, *De Motu*. Here we quote from its English translation [57, pp. 209-227]. A good discussion of Berkeley's philosophy of motion can be found elsewhere ([58], [59] and [60]).

In Section 112 of the *Principles* he outlined a relational theory, as follows, [61]:

> 112. But, notwithstanding what has been said, I must confess it does not appear to me that there can be any motion other than *relative*; so that to conceive motion there must be at least conceived two bodies, whereof the distance or position in regard to each other is varied. Hence, if there was one only body in being it could not possibly be moved. This seems evident, in that the idea I have of motion doth necessarily include relation.

Analogously, in Section 63 of *De Motu* we read, [57]:

> 63 No motion can be recognized or measured, unless through sensible things. Since then absolute space in no way affects the senses, it must necessarily be quite useless for the distinguishing of motions. Besides, determination or direction is essential to motion; but that consists in relation. Therefore it is impossible that absolute motion should be conceived.

But Berkeley also seems to contradict himself, as did Leibniz, when he takes the forces into account. He also gives forces an absolute reality, and in this way is led astray by the Newtonian arguments. For instance, in paragraph 113 of the *Principles*, he writes:

> 113. But, though in every motion it be necessary to conceive more bodies than one, yet it may be that one only is moved, namely,

that on which the force causing the change in distance or situation
of the bodies, is impressed. For, however some may define relative
motion, so as to term that body *moved* which changes its distance
from some other body, whether the force or action causing that
change were impressed on it or no, yet as relative motion is that
which is perceived by sense, and regarded in the ordinary affairs of
life, it should seem that every man of common sense knows what
it is as well as the best philosopher. Now, I ask any one whether,
in his sense of motion as he walks along the streets, the stones he
passes over may be said to *move*, because they change distance with
his feet? To me it appears that though motion includes a relation
of one thing to another, yet it is not necessary that each term of the
relation be denominated from it. As a man may think of somewhat
which does not think, so a body may be moved to or from another
body which is not therefore itself in motion.

But even if there is only relative motion, how could he explain Newton's
bucket and two globes experiments without introducing absolute space? He is
not completely clear on this, but he seems to have meant that the concave form
of the water in the spinning bucket only appeared due to its relative rotation
with respect to the set of fixed stars. And the same would explain the tension
of the string in the two globes experiment. These dynamical effects would be
related to the kinematical motion between the test body and the stars. They
would not be related to a motion of the test body relative to absolute space. In
order to show this possible interpretation of Berkeley's ideas, we present here
Section 114 of the *Principles* where he discusses Newton's bucket experiment:

114. As the place happens to be variously defined, the motion which
is related to it varies. A man in a ship may be said to be quiescent
with relation to the sides of the vessel, and yet move with relation
to the land. Or he may move eastward in respect of the one, and
westward in respect to the other. In the common affairs of life men
never go beyond the earth to define the place of any body; and
what is quiescent in respect of that is accounted *absolutely* to be so.
But philosophers, who have a greater extent of thought, and juster
notions of the system of things, discover even the earth itself to be
moved. In order therefore to fix their notions they seem to conceive
the corporeal world as finite, and the utmost unmoved walls or shell
thereof to be the place whereby they estimate true motions. If we
sound our conceptions, I believe we may find all the absolute motion
we can frame an idea of to be at bottom no other than relative
motion thus defined. For, as hath been already observed, absolute

motion, exclusive of all external relations, is incomprehensible; and to this kind of relative motion all the above-mentioned properties, causes, and effects ascribed to absolute motion will, if I mistake not, be found to agree. As to what is said of the centrifugal force, that it does not at all belong to circular relative motion, I do not see how this follows from the experiment which is brought to prove it. See *Philosophiae Naturalis Principia Mathematica, in Schol. Def. VIII.* For the water in the vessel at that time wherein it is said to have the greatest relative circular motion, hath, I think, no motion at all; as is plain from the foregoing section.

When he says that philosophers "conceive the corporeal world as finite, and the utmost unmoved walls or shell thereof to be the place whereby they estimate true motions," he means the set of fixed stars. According to Berkeley the philosophers put the set of stars at rest by convention and establish motion of other celestial bodies relative to this frame of reference of the fixed stars. When Berkeley writes that in the beginning of Newton's bucket experiment the "water in the vessel has no motion at all," he presumably means no motion of the water relative to the earth or relative to the set of stars. After all, in the situation described by Newton there is the greatest relative circular motion between the bucket and the water after the bucket was released and spun fastest relative to the earth, while the water did not yet have time to rotate together with the bucket. If this is the case, it would follow that to Berkeley the concave of the water only appears when there is a relative rotation between the water and the earth (or between the water and the set of stars), although we cannot ascribe a real or absolute rotation to the water or to the earth (not even to the set of stars). But obviously here we are ascribing more to Berkeley than what he really wrote. As we saw before when discussing his §113 (see especially the first sentence of this paragraph), Berkeley is sometimes confused by Newton's arguments. In these cases he speaks of the forces as something absolute, asserting that we can determine which body is really and absolutely in motion by observing on which body the force is acting. However, this is meaningless in a truly relational theory.

He suggested replacing Newton's absolute space by the set of fixed stars (more clearly than Leibniz) in Section 64 of *De Motu*, [57]:

64 Further, since the motion of the same body may vary with the diversity of relative place, nay actually since a thing can be said in one respect to be in motion and in another respect to be at rest, to determine true motion and true rest, for the removal of ambiguity and for the furthearance of the mechanics of these philosophers who take the wider view of the system of things, it would be enough to

bring in, instead of absolute space, relative space as confined to the heavens of the fixed stars, considered as at rest. But motion and rest marked out by such relative space can conveniently be substituted in place of the absolutes, which cannot be distinguished from them by any mark. (...)

Two hundred years later, Mach would also propose replacing Newton's absolute space with the set of fixed stars.

In Sections 58 to 60 of *De Motu* Berkeley discussed Newton's two globes and experiments as follows, [57]:

58 From the foregoing it is clear that we ought not to define the true place of the body as the part of absolute space which the body occupies, and true or absolute motion as the change of true or absolute place; for all place is relative just as all motion is relative. But to make this appear more clearly we must point out that no motion can be understood without some determination or direction, which in turn cannot be understood unless besides the body in motion our own body also, or some other body, be understood to exist at the same time. For *up, down, left,* and *right* and all places and regions are founded in some relation, and necessarily connote and suppose a body different from the body moved. So that if we suppose the other bodies were annihilated and, for example, a globe were to exist alone, no motion could be conceived in it; so necessarily is it that another body should be given by whose situation the motion should be understood to be determined. The truth of this opinion will be very clearly seen if we shall have carried out thoroughly the supposed annihilation of all bodies, our own and that of others, except that solitary globe.

59 Then let two globes be conceived to exist and nothing corporeal besides them. Let forces then be conceived to be applied in some way; whatever we may understand by the application of forces, a circular motion of the two globes round a common centre cannot be conceived by the imagination. Then let us suppose that the sky of the fixed stars is created; suddenly from the conception of the approach of the globes to different parts of that sky the motion will be conceived. That is to say that since motion is relative in its own nature, it could not be conceived before the correlated bodies were given. Similarly no other relation can be conceived without correlates.

60 As regards circular motion many think that, as motion trully circular increases, the body necessarily tends ever more and more

away from its axis. This belief arises from the fact that circular motion can be seen taking its origin, as it were, at every moment from two directions, one along the radius and the other along the tangent, and if in this latter direction only the impetus be increased, then the body in motion will retire from the centre, and its orbit will cease to be circular. But if the forces be increased equally in both directions the motion will remain circular though accelerated - which will not argue an increase in the forces of retirement from the axis, any more than in the forces of approach to it. Therefore we must say that the water forced round in the bucket rises to the sides of the vessel, because when new forces are applied in the direction of the tangent to any particle of water, in the same instant new equal centripetal forces are not applied. From which experiment it in no way follows that absolute circular motion is necessarily recognized by the forces of retirement from the axis of motion. Again, how those terms *corporeal forces* and *connation* are to be understood is more than sufficiently shown in the foregoing discussion.

In other words, for Berkeley it is only meaningful to state that the two globes rotate when we have other bodies to refer motion to. Moreover, this rotation will be only relative, as we cannot say if the globes are rotating while the sky of fixed stars is at rest, or *vice versa*. But he does not say explicitly that the tension in the string connecting the two globes will appear only when there is relative rotation between the globes and the stars. Nor does he say explicitly that the tension in the string will only appear when the stars are created, as was pointed out by Clarke.

As regards his discussion of the bucket experiment, once again Berkeley did not emphasize the role of the fixed stars in generating the centrifugal forces. Nor did he say that the water would be flat if the other bodies in the universe were annihilated.

These aspects were pointed out clearly by Jammer ([33, p. 109]): "Berkeley's statement obviously cannot be considered as being equivalent to what is called in modern cosmology 'Mach's principle' (that is, that the inertia of any body is determined by the masses of the universe and their distribution), as Berkeley confines himself to the problem of the perception and comprehensibility of motion and ignores in this context the dynamical aspect of motion."

But even if this was the correct interpretation of his ideas, Berkeley did not implement them quantitatively. He did not present a specific force law to show that when we keep the globes at rest and rotate the set of stars there appears a real centrifugal force creating a tension in the string due to this relative rotation.

Finally, he did not mention the proportionality between inertial and gravitational masses. He did not suggest the possibility that the centrifugal force might be due to a *gravitational* interaction with distant matter.

Many other authors discussed these aspects of Newtonian theory before Mach, *e.g.* Euler, d'Alembert, Kant, *etc.* However, they did not advance much further beyond Newton, Leibniz or Berkeley. Most of them defended Newton's ideas. A short summary can be found elsewhere ([33, Chapter 5], [54, pp. xl to xlix] and [62]). We will not enter into more details here, as the main ideas were developed by Leibniz and Berkeley. Later on these ideas were greatly extended and explored further by Mach. This is the subject of the next chapter.

Chapter 6

Mach and Newton's Mechanics

6.1   Inertial Frame of Reference

In this chapter we present the criticisms made by Ernst Mach (1838-1916) of Newtonian mechanics. We will try to follow some of the examples discussed in the previous chapters to illustrate the shortcomings of classical mechanics according to Mach, and show how he suggested overcoming them.

For a biography of Mach, see reference [63].

We begin with the the problem of uniform rectilinear motion. According to Newton's first law of motion (the law of inertia) if there is no net force acting on a body it will stay at rest or will move along a straight line with constant velocity. But relative to what frame of reference will the body stay at rest or move with a constant velocity? According to Newton, the motion is relative to absolute space or any other frame which moves with a constant velocity relative to absolute space. The problem with this statement is that we do not have any access to absolute space. We cannot know our position or velocity relative to absolute space. Mach wanted to get rid of the notions of absolute space and time. In the Preface to the first German edition (1883) of his book *The Science of Mechanics*, Mach wrote: "The present volume is not a treatise upon the application of the principles of mechanics. Its aim is to clear up ideas, expose the real significance of the matter, and get rid of metaphysical obscurities" [39]. In the Preface of the seventh German edition (1912) of this book he wrote:

> The character of the book has remained the same. With respect to
> the monstrous conceptions of absolute space and absolute time I can

retract nothing. Here I have only shown more clearly than hitherto that Newton indeed spoke much about these things, but throughout made no serious application of them. His fifth corollary[1] contains the only practically usable (probably approximate) *inertial system.*

What did Mach suggest as an alternative to Newton's absolute space? He proposed the fixed stars and the rest of matter in the universe [39, pp. 285-6]:

> The comportment of terrestrial bodies with respect to the earth is reducible to the comportment of the earth with respect to the remote heavenly bodies. If we were to assert that we knew more of moving objects than this their last-mentioned, experimentally-given comportment with respect to the celestial bodies, we should render ourselves culpable of a falsity. When, accordingly, we say, that a body preserves unchanged its direction and velocity in space, our assertion is nothing more or less than an abbreviated reference to the entire universe.

His clearest answer appears in pages 336-7 of this book, our emphasis:

> 4. I have now another important point to discuss in opposition to C. Neumann,[2] whose well-known publication on this topic preceded mine[3] shortly. I contended that the direction and velocity which is taken into account in the law of inertia had no comprehensible meaning if the law was referred to "absolute space." As a matter of fact, we can metrically determine direction and velocity only in a space of which the points are marked directly or indirectly by given bodies. Neumann's treatise and my own were successful in directing attention anew to this point, which had already caused Newton and Euler much intellectual discomfort; yet nothing more than partial attempts at solution, like that of Streintz, have resulted. *I have remained to the present day the only one who insists upon referring the law of inertia to the earth, and in the case of motions of great spatial and temporal extent, to the fixed stars.*

It is difficult to disagree with Mach on this point. This last sentence is much more practical than Newton's formulation of the first law in terms of absolute space. In typical laboratory experiments (such as the study of springs, collision of two billiards balls, *etc.*), which last much less than one hour and which do

---

[1] *Principia*, 1687, p. 19.

[2] *Die Principien der Galilei-Newton'schen Theorie*, Leipzig, 1870.

[3] *Erhaltung der Arbeit*, Prague, 1872. (Translated in part in the article on "The Conservation of Energy," *Popular Scientific Lectures*, third edition, Chicago, 1898.)

not extend very far in space compared to the earth's radius, we can utilize the earth as our inertial system. Consequently we can apply Newton's laws of motion without fictitious forces in this frame in order to study these motions with reasonable accuracy. On the other hand, in experiments which last many hours, such as Foucault's pendulum, or in which we study motions with long space and time scales, such as the winds, oceanic currents *etc.*, a better inertial frame than the earth is the frame defined by the stars. The fixed stars are also a good inertial frame for studying the rotation of the earth each day, its flattening at the poles or its translation around the sun in one year. Nowadays, we might say that a better inertial frame for studying the rotation or motion of the galaxy as a whole (relative to other galaxies, for instance) is the frame of reference defined by the external galaxies or the frame of reference in which the cosmic background radiation is isotropic.

6.2 Absolute Time

Mach also rejected Newton's absolute time. His points of view as regards time were presented clearly on p. 273 of *The Science of Mechanics*:

> It is utterly beyond our power to *measure* the changes of things by *time*. Quite the contrary, time is an abstraction, at which we arrive by means of the changes of things; made because we are not restricted to any one *definite* measure, all being interconnected. A motion is termed uniform in which equal increments of space described correspond to equal increments of space described by some motion with which we form a comparison, as the rotation of the earth. A motion may, with respect to another motion, be uniform. But the question whether a motion is *in itself* uniform, is senseless. With just as little justice, also, may we speak of an "absolute time"—*of a time independent of* change. This absolute time can be measured by comparison with no motion; it has therefore neither a practical nor a scientific value; and no one is justified in saying that he knows aught about it. It is an idle metaphysical conception.

Mach thought that we could replace the time t which appears in Newton's laws of motion by the angle of rotation of the Earth with respect to the fixed stars. He expressed this view on pages 287 and 295 of *The Science of Mechanics*:

> When we reflect that the time-factor that enters into the acceleration is nothing more than a quantity that is the measure of the distances (or angles of rotation) of the bodies of the universe, we see that even in the simplest case, in which apparently we deal with the mutual

action of only *two* masses, the neglecting of the rest of the world is *impossible*. (...)

We measure time by the angle of rotation of the earth, but could measure it just as well by the angle of rotation of any other planet. But, on that account, we would not believe that the *temporal* course of all physical phenomena would have to be disturbed if the earth or the distant planet referred to should suddenly experience an abrupt variation of angular velocity. We consider the dependence as not immediate, and consequently the temporal orientation as *external*. Nobody would believe that the chance disturbance - say by an impact - of one body in a system of uninfluenced bodies which are left to themselves and move uniformly in a straight line, where all the bodies combine to fix the system of coordinates, will immediately cause a disturbance of the others as a consequence. The orientation is external here also. Although we must be very thankful for this, especially when it is purified from meaninglessness, still the natural investigator must feel the need of further insight - of knowledge of the *immediate* connections, say, of the masses of the universe. There will hover before him as an ideal an insight into the principles of the whole matter, from which accelerated and inertial motions result in the *same* way. (...)

Once more it is difficult to disagree with Mach on these points of view.

6.3   The Two Rotations of the Earth

Mach was aware of the observational evidence that the kinematical rotation of the earth relative to the fixed stars is the same as the dynamical rotation of the earth relative to inertial frames. In other words, the best inertial system of reference known to us (in which we can apply Newton's second law of motion without centrifugal, Coriolis or other fictitious forces) does not rotate relative to the set of fixed stars. He expressed this on pages 292-3 of *The Science of Mechanics*: "Seeliger has attempted to determine the relation of the inertial system to the empirical astronomical system of coordinates which is in use, and believes that he can say that the empirical system cannot rotate about the inertial system by more than some seconds of arc in a century." Seeliger's work of 1906 has been discussed by Jammer [33, p. 141].

Nowadays we know that if there is a rotation between these two frames it is smaller than 0.4 seconds of arc per century [64], or:

$$\omega_k - \omega_d \leq \pm 0.4 \ \text{sec/century} = \pm 1.9 \times 10^{-8} \ \text{rad/year} .$$

As $\omega_k = 2\pi/T = 2\pi/(24 \text{ hours}) = 7.29 \times 10^{-5}$ rad/s we get

$$\frac{\omega_k - \omega_d}{\omega_k} \leq \pm 8 \times 10^{-12} \ .$$

Few facts in physics have a precision of one part in 10^{11}, as we find here (another example, as we saw previously, is the proportionality between inertia and weight). This is one of the strongest empirical supports for Mach's principle. It is difficult to accept this fact as a simple coincidence. As we have seen, it is equivalent to the statement that the universe as a whole (the set of distant galaxies) does not rotate relative to absolute space. It suggests that distant matter determines and establishes the best inertial frame. If this is the case, we need to understand and explain this connection between distant matter and local inertial systems. No answer to this puzzle is to be found in Newtonian mechanics as in it there is no relation between the stars and inertial frames.

6.4 Inertial Mass

Another problem in classical mechanics is the notion of inertial mass, the mass which appears in Newton's second law of motion, in the linear momentum and in the kinetic energy. Newton defined it as the product of the volume of the body by its density. This is a poor definition, as we usually define the density by the ratio of the inertial mass (or quantity of matter) and volume of a body. Newton's definition would only be useful (and would only avoid vicious circles) if Newton had specified previously how to define and measure the density of a body without utilizing the mass concept, but he did not do this. The first article written by Mach where he criticized this definition and presented a better one is from 1868. It was reprinted in Mach's book *The History and the Root of the Principle of the Conservation of Energy* of 1872 [65, pp. 80-85]. In *The Science of Mechanics* he elaborated further his new proposal and wrote: "Definition I is, as has already been set forth, a pseudo-definition. The concept of mass is not made clearer by describing mass as the product of the volume into the density, as density itself denotes simply the mass of unit of volume. The true definition of mass can be deduced only from the dynamical relations of bodies" [39], p. 300.

Instead of Newton's definition, Mach proposed the following [39, p. 266]:

> *All those bodies are bodies of equal mass, which, mutually acting on each other, produce in each other equal and opposite accelerations.*

> We have, in this, simply designated, or named, an actual relation of things. In the general case we proceed similarly. The bodies A and B receive respectively as the result of their mutual action

(see Figure) the accelerations $-\varphi$ and $+\varphi'$, where the senses of the accelerations are indicated by the signs.

We say then, B has φ/φ' times the mass of A. If we take A as our unit, we assign to that body the mass m which imparts to A m times the acceleration that A in the reaction imparts to it. The ratio of the masses is the negative inverse ratio of the counter-accelerations. That these accelerations always have opposite signs, that there are therefore, by our definition, only positive masses, is a point that experience teaches, and experience alone can teach. In our concept of mass no theory is involved; "quantity of matter" is wholly unnecessary in it; all it contains is the exact establishment, designation, and determination of a fact.

In this key definition of inertial mass, Mach did not specify clearly the frame of reference with respect to which the accelerations should be measured. It is simple to see that this definition depends on the frame of reference. For instance, observers in two frames which are accelerated relative to one another will find different mass ratios by analysing the same interaction of two bodies if each observer utilizes his own frame of reference to define the accelerations and arrive at the masses. Let us give an example. We consider a one-dimensional problem in which two bodies, 1 and 2, interacting with one another obtain accelerations a_1 and $-a_2$ relative to a frame of reference O, as in Figure 6.1.

Now suppose a frame of reference O' has an acceleration $a_{o'}$ relative to O, along the direction of the accelerations of 1 and 2. The accelerations of bodies 1 and 2 relative to O' will be given by, respectively: $a_1' = a_1 - a_{o'}$ and $a_2' = -a_2 - a_{o'}$, as in Figure 6.2.

Utilizing Mach's definition, the mass-ratio of bodies 1 and 2 relative to O is given by $m_1/m_2 = -(-a_2/a_1) = a_2/a_1$. On the other hand, their mass-ratio relative to O' is found to be: $m_1'/m_2' = -(-a_2'/a_1') = (a_2 + a_{o'})/(a_1 - a_{o'}) \neq m_1/m_2$. In other words, if we can utilize any frame of reference to define the mass-ratios, then this definition becomes meaningless. After all, there will be as many different mass-ratios as there are frames of reference accelerated relative to one another. The value of m_1/m_2 would depend on the system of reference, and this is certainly undesirable.

But it is evident from his writings that Mach had in mind the frame of fixed stars as the only frame to be utilized in this definition. This was shown conclusively in an important paper by Yourgrau and van der Merwe [66]. We may cite a few passages from Mach to prove this point. When discussing Newton's bucket experiment, Mach writes: "The natural system of reference is for him [Newton] that which has any uniform motion or translation without rotation (relatively to the sphere of fixed stars)" [39], p. 281. The words in parenthesis are Mach's, and did not come from Newton. On page 285 he writes: "Now, in order to have a generally valid system of reference, Newton ventured the fifth corollary of the *Principia* (p. 19 of the first edition). He imagined a momentary terrestrial system of coordinates, for which the law of inertia is valid, held fast in space without any rotation relatively to the fixed stars." Once more, these last words (relatively to the fixed stars) are from Mach, as Newton did not mention the fixed stars in his fifth corollary. On pages 294-5 Mach wrote: "There is, I think, no difference of meaning between Lange and myself (...) about the fact that, at the present time, the set of the fixed stars is the only practically usable system of reference, and about the method of obtaining a new system of reference by gradual correction."

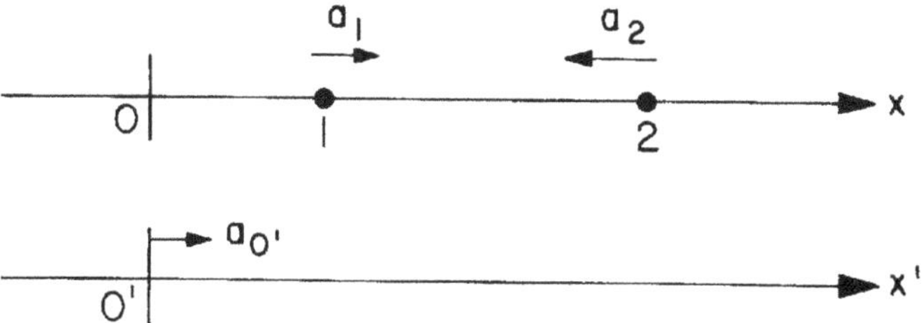

Figure 6.1: Accelerations of two bodies relative to O.

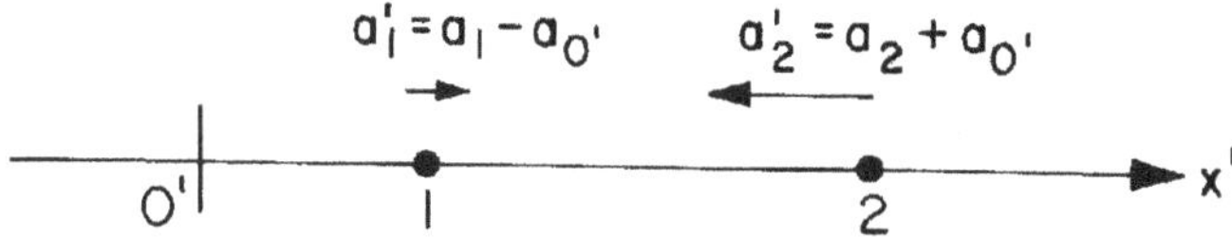

Figure 6.2: Acceleration of the bodies relative to O'.

It should be observed that nowadays the accepted definition of inertial mass is Mach's ($m_1/m_2 = -a_2/a_1$) and not Newton's ($m = \rho V$). See, for instance, Symon's book [27, Section 1.3]. Despite this fact, Mach's name is not usually quoted in this connection.

Mach's operational definition of inertial mass is one of his main contributions to the foundations of classical mechanics.

6.5   Mach's Formulation of Mechanics

After clarifying these points, we now present here Mach's own formulation of mechanics, which he suggested in order to replace Newton's postulates and corollaries. Mach presented this formulation for the first time in 1868 [65, especially pp. 84-5]. Here we present his final formulation [39, pp. 303-4]:

> Even if we adhere absolutely to the Newtonian points of view, and disregard the complications and indefinite features mentioned, which are not removed but merely concealed by the abbreviated designations "Time" and "Space," it is possible to replace Newton's enunciations by much more simple, methodically better arranged, and more satisfactory propositions. Such, in our estimation, would be the following:
>
> a. *Experimental proposition.* Bodies set opposite each other induce in each other, under certain circumstances to be specified by experimental physics, contrary *accelerations* in the direction of their line of junction. (The principle of inertia is included in this.)
>
> b. *Definition.* The mass-ratio of any two bodies is the negative inverse ratio of the mutually induced accelerations of those bodies.
>
> c. *Experimental Proposition.* The mass-ratios of bodies are independent of the character of the physical states (of the bodies) that condition the mutual accelerations produced, be those states electrical, magnetic, or what not; and they remain, moreover, the same, whether they are mediated or immediately arrived at.
>
> d. *Experimental Proposition.* The accelerations which any number of bodies A, B, C ... induce in a body K, are independent of each other. (The principle of the parallelogram of forces follows immediately from this.)
>
> e. *Definition.* Moving force is the product of the mass-value of a body into the acceleration induced in that body.

These are clear and reasonable propositions, provided we understand the frame of reference to which the accelerations are to be referred. As we have seen, to Mach a reasonable frame of reference for these accelerations was the earth. If we need a greater precision and more accurate mass-ratios, then according to Mach we need to utilize the frame of fixed stars.

This Machian formulation of mechanics is vastly superior than the Newtonian one. After all it is based only on practical procedures and on facts of experience, without metaphysical concepts such as absolute space and time. However, this is not enough. It does not explain the proportionality between inertia and weight (or between m_i and m_g), it does not explain why the fixed stars are a good inertial system (or why the set of fixed stars does not rotate relative to inertial systems), nor does it explain the origin of fictitious forces (such as the centrifugal and Coriolis forces). Although it represents a tremendous progress compared with Newton, Leibniz and Berkeley, a complete quantitative implementation of relational mechanics requires much more than Mach accomplished. Neverthless, he took a large step in the right direction.

6.6 Relational Mechanics

Beyond these clarifications and important new formulation, Mach presented two extremely relevant suggestions and insights. The first was to emphasize that in physics which should have only relational quantities. That is, physics should depend only on relative distance between bodies and their relative motions. No absolute positions and velocities should appear in the theory as they do not appear in the experiments. His second suggestion was related to Newton's bucket experiment. We discuss this in the next section, analysing first his remarks about relational mechanics.

Mach's statements in this regard can be found at several places in *The Science of Mechanics*, from which we quote the following (our emphasis):

> p. 279: If, in a material spatial system, there are masses with different velocities, which can enter into mutual relations with one another, these masses present to us forces. We can only decide how great these forces are when we know the velocities to which those masses are to be brought. *Resting* masses too are forces if *all* the masses do not rest. Think, for example, of Newton's rotating bucket in which the water is not yet rotating. If the mass m has the velocity v_1 and it is to be brought to the velocity v_2, the force which is to be spent on it is $p = m(v_1 - v_2)/t$, or the work which is to be expended is $ps = m(v_1^2 - v_2^2)$. *All masses and all velocities, and consequently all forces, are relative.* There is no decision about relative and absolute which we can possibly meet, to which we are forced, or from which we can obtain any intellectual or other advantage. When quite modern authors let themselves be led astray by the Newtonian arguments which are derived from the bucket of water, to distinguish between relative and absolute motion, they do not reflect that the system of the world is only given *once* to us, and the Ptolemaic or Copernican view is *our* interpretation, but both are equally actual. *Try to fix Newton's bucket and rotate the heaven of fixed stars and then prove the absence of centrifugal forces.*

pp. 283-4: Let us now examine the point on which Newton, apparently with sound reasons, rests his distinction of absolute and relative motion. If the earth is affected with an *absolute* rotation about its axis, centrifugal forces are set up in the earth: it assumes an oblate form, the acceleration of gravity is diminished at the equator, the plane of Foucault's pendulum rotates, and so on. All these phenomena disappear if the earth is at rest and the other heavenly bodies are affected with absolute motion round it, such that the same *relative* rotation is produced. This is, indeed, the case, if we start ab initio from the idea of absolute space. But if we take our stand on the basis of facts, we shall find we have knowledge only of *relative* spaces and motions. *Relatively, not considering the unknown and neglected medium of space, the motions of the universe are the same whether we adopt the Ptolemaic or Copernican mode of view. Both views are, indeed, equally correct; only the latter is more simple and more practical.* The universe is not *twice* given, with an earth at rest and an earth in motion; but only *once*, with its *relative* motions, alone determinable. It is, accordingly, not permitted us to say how things would be if the earth did not rotate. We may interpret the one case that is given to us, in different ways. If, however, we so interpret it that we come into conflict with experience, our interpretation is simply wrong. *The principles of mechanics can, indeed, be so conceived, that even for relative rotations centrifugal forces arise.*

From these and other quotations we understand that a relational mechanics following Mach's point of view should depend only on relative quantities, in other words, on the distance between the bodies, $r_{mn} = |\vec{r}_m - \vec{r}_n|$, and their time derivatives: $\dot{r}_{mn} = dr_{mn}/dt$, $\ddot{r}_{mn} = d^2r_{mn}/dt^2$, d^3r_{mn}/dt^3, *etc.* Moreover, the concepts of absolute space and time should not appear.

6.7   Mach and the Bucket Experiment

When Mach discussed Newton's bucket experiment, he emphasized the fact that we cannot neglect the heavenly bodies in the analysis. According to Mach the parabolic shape of the spinning water is due to its rotation relative to the fixed stars, and not due to its rotation relative to absolute space. For instance, on p. 284 of *The Science of Mechanics* [39], he wrote:

Newton's experiment with the rotating vessel of water simply informs us, that the relative rotation of the water with respect to the sides of the vessel produces *no* noticeable centrifugal forces, but

that such forces *are* produced by its relative rotation with respect
to the mass of the earth and the other celestial bodies. No one is
competent to say how the experiment would turn out if the sides of
the vessel increased in thickness and mass till they were ultimately
several leagues thick. The one experiment only lies before us, and
our business is, to bring it into accord with the other facts known
to us, and not with the arbitrary fictions of our imagination.

The most important point here is that the issue is not just one of language.
Instead of Newton's absolute space we could speak of Mach's frame of fixed
stars, and then all would be settled. This would be the case if it were only a
question of language. But the passages cited above indicate a deepest meaning.
In fact they suggest a dynamical origin for the centrifugal force, that Mach
sought the centrifugal force as a real force which appears in a frame of reference
in which the sky of stars is rotating. This cannot be derived from Newton's
laws of motion, nor even from his universal law of gravitation.

Newton's experiment is represented in Figure 6.3. The bucket and the water
are rotating together with angular velocity $\omega\hat{z}$ relative to the earth and to the
fixed stars. The surface of the water is concave. We choose the z axis along
the axis of the bucket, which does not need to be along the north-south axis.
The rotation of the bucket and water relative to the earth are much greater
than the diurnal rotation of the earth relative to the fixed stars. Thus we can
consider the earth to be essentially without acceleration relative to the frame
of fixed stars during this experiment.

We can distinguish clearly Newton's point of view from Mach's with Figures
6.4 and 6.5.

In Figure 6.4 we assume that the bucket, water and earth are at rest relative
to absolute space and that the set of stars rotate relative to this frame or to
the earth with an angular velocity $-\omega\hat{z}$. According to Newton the water will
remain flat, as it is at rest relative to absolute space and the spinning set of
stars exert no net gravitational force on the water molecules.

In Figure 6.5 we have the outcome of this thought experiment according to
Mach. Provided the relative rotation is the same as in Newton's original and
real experiment (rotating the bucket relative to the earth and relative to the
set of fixed stars with $+\omega\hat{z}$), the surface of the water should remain concave.
To Mach absolute space does not exist and cannot play any role here. Only the
relative rotation between the water and the fixed stars should matter. If the
kinematical situation is the same (stars at rest relative to an arbitrary frame
of reference and water spinning with $+\omega\hat{z}$, or water at rest relative to another
frame of reference and the stars spinning with $-\omega\hat{z}$), then the dynamical effects
must also be the same (the water must rise towards the sides of the vessel
in both cases). The only thing Mach did not know is that the cause of the

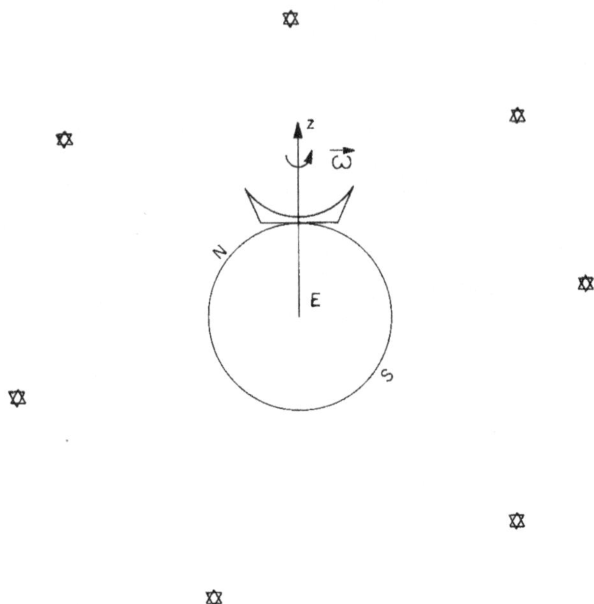

Figure 6.3: Newton's bucket experiment.

concavity of the water surface is its rotation relative to distant galaxies and not relative to the fixed stars. Later on we explain why.

Obviously the situations of Figures 6.4 and 6.5 are not completely equivalent to Newton's real experiment. The kinematical equivalence would be complete only if the earth rotated together with the fixed stars with $-\omega\hat{z}$ relative to the bucket and water. But here we are neglecting the tangential forces exerted by the spinning earth on the molecules of water. We are assuming that the force exerted by the earth on the water is essentially its weight pointing downwards, regardless of the rotation of the earth relative to the water.

Mach wrote: "try to fix Newton's bucket and rotate the heaven of fixed stars and then prove the absence of centrifugal forces." The chief importance of this statement was that it implied clearly that the centrifugal force is due to the relative rotation between bodies which experience these forces and the distant masses in the universe. Many physicists have been heavily influenced by Mach's writings, which were more influential than the similar ideas presented earlier by Leibniz and Berkeley.

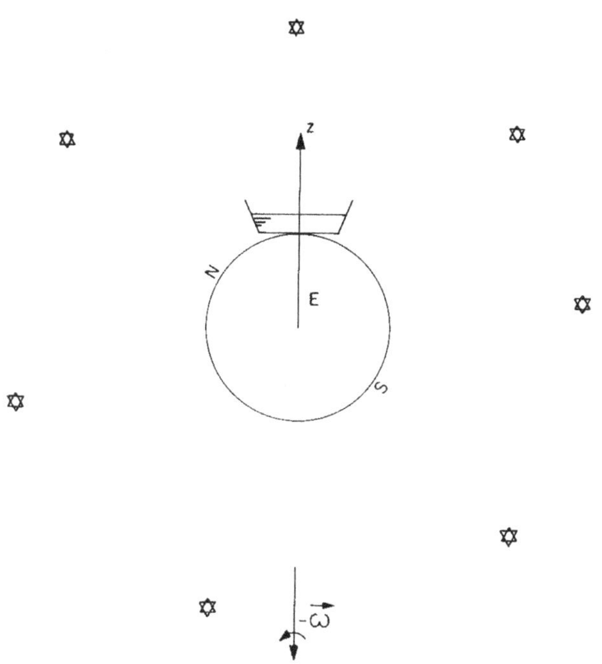

Figure 6.4: What should be expected in Newton's theory if we fixed the bucket relative to the earth and rotated the set of fixed stars.

6.8 Mach's Principle

Mach did not specify clearly in his writings any specific principle which he advocated. Despite this fact, he presented cogently argued points of view against Newton's absolute space and time, in favour of a relational physics, in favour of the physical reality of the fictitious forces, supposed that Newton's bucket experiment showed a connection between the curvature of the water and its rotation relative to the fixed stars, *etc.* These ideas became generally known by the name "Mach's Principle." Here we discuss how different authors have used this principle [67].

The first to utilize the expressions "Mach's principle" and "Mach's postulates" was M. Schlick in 1915 [68, see especially pp. 10 and 47, note 2]. Apparently he was referring to Mach's general proposals of a relativity of all motions (there were no motions relative to space, but only motion of matter relative to other matter). According to Schlick, a consequence of this proposal is that "the cause of inertia must be assumed to be an interaction of masses."

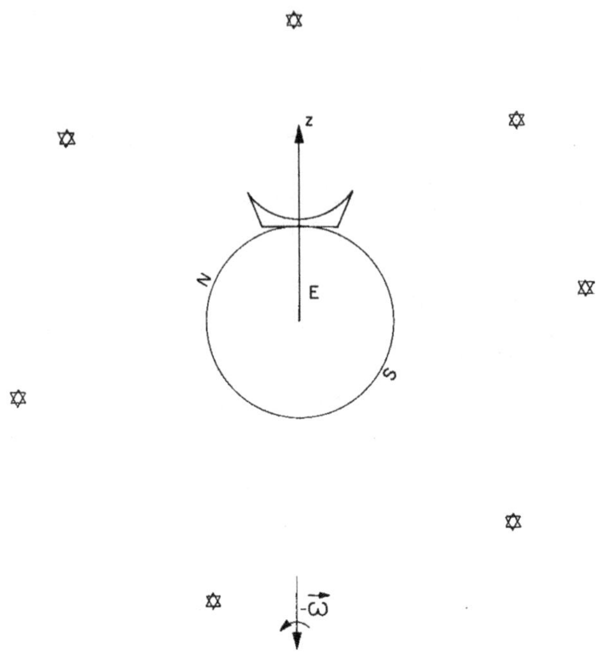

Figure 6.5: What should be expected in Mach's theory if we fixed the bucket relative to the earth and rotated the set of fixed stars.

The expression became widely known and utilized after Einstein's article of 1918 on this topic. In this article he says the following concerning his general theory of relativity (see [69, pp. 185-186] for this English translation):

The theory, as it now appears to me, rests on three main points of view, which, however, are by no means independent of each other...:

a) *Relativity principle:* The laws of nature are merely statements about space-time coincidences; they therefore find their only natural expression in generally covariant equations.

b) *Equivalence principle:* Inertia and weight are identical in nature. It follows necessarily from this and from the result of the special theory of relativity that the symmetric 'fundamental tensor' $[g_{\mu\nu}]$ determines the metrical properties of space, the inertial behavior of bodies in it, as well as gravitational effects. We shall denote the state of space described by the fundamental tensor as the 'G-field.'

c) *Mach's Principle*[4]: The G-field is *completely* determined by the masses of the bodies. Since mass and energy are identical in accordance with the results of the special theory of relativity and the energy is described formally by means of the symmetric energy tensor $(T_{\mu\nu})$, this means that the G-field is conditioned and determined by the energy tensor of the matter.

Different formulations of Mach's principle, as given by Einstein, have been pointed out by J. Barbour in [69, p. 179].

Below are the words of some other authors when referring to this principle:

Sciama: "Inertial frames are those which are unaccelerated relative to the 'fixed stars', that is, relative to a suitably defined mean of all the matter in the universe" [70].

Brown: "Inertia is not due to movement with respect to 'absolute space', but due to surrounding matter" [71].

Kaempffer: "By 'Mach's Program' is meant the intention to understand all inertial effects as being caused by gravitational interaction" [72].

Moon and Spencer: "Inertia is not an inherent property of matter but is the result of forces caused by the distant galaxies" [73].

Schiff: "The inertial properties of matter on the local scene derive in some way from the existence of the distant masses of the universe and their distribution in space" [64].

Bunge: "The motion and consequently the mass of every single body is determined (caused, produced) by the remaining bodies in the universe" [74].

Jammer: "The inertia of any body is determined by the masses of the universe and their disbribution" [33, p. 109].

Reinhardt: "The inertial mass of a body is caused by its interaction with the other bodies in the universe" [75].

Phipps: "When the subway jerks, it's the fixed stars that throw you down," Phipps says that this raw form was attributed by P. Frank to Mach himself [76].

Raine: "Inertial forces should be generated entirely by the motion of a body relative to the bulk of matter in the universe" [77].

Barbour: "Mach suggested that inertial motion here on the earth and in the solar system is causally determined in accordance with some quite definite but as yet unknown law by the totality of the matter in the universe" [4].

As Mach himself did not specify an explicit principle but only general ideas as presented above, we utilize in this work these ideas as "Mach's Principle."

---

[4]Hitherto I [Einstein] have not distinguished between principles (a) and (c), and this was confusing. I have chosen the name 'Mach's principle' because this principle has the significance of a generalization of Mach's requirement that inertia should be derived from an interaction of bodies.

6.9   What Mach did not Show

Here we present briefly some results which are embodied in Mach's principle but that he did not implement quantitatively.

Mach did not emphasize that the inertia of a body should be due to its *gravitational* interaction with other bodies in the universe. By inertia we mean here the inertial mass of the body and its inertial properties to resist being accelerated. In principle this connection between the inertia of a body and the distant celestial bodies might be due to any kind of interaction already known (electrical, magnetic, elastic, *etc.*) or even to a new kind of interaction. In no place did he state that the inertia of a body should come from its gravitational interaction with the fixed stars. The first to suggest this seem to have been the Friedlaender brothers in 1896 [78]. Their work has been partially translated to English recently [79]. This idea was also adopted by Höfler in 1900, by W. Hofmann in 1904 (partial English translation: [80]), by Einstein in 1912 (partial English translation of the relevant section: [69, p. 180]), by Reissner in 1914-1915 (English translations: [81] and [82]), by Schrödinger in 1925 (English translation: [83]) and by many other authors ever since ([12, Sections 7.6 and 7.7] and [68]). In chapter 11 we discuss all of these fonts in more detail.

Nor did Mach derive the proportionality between inertial and gravitational masses. On page 270 of *The Science of Mechanics* Mach wrote the following: "The fact that *mass* can be *measured* by *weight*, where the acceleration of gravity is invariable, can also be deduced from our definition of mass." This deduction is not warranted. The fact that two bodies of different mass (and/or chemical composition, and/or form, *etc.*) fall to the earth with the same acceleration in vacuum cannot be derived from Mach's definition of mass, but is shown only from experience. We might let two bodies A and B on a frictionless table interact through a spring and determine their mass ratio by Mach's definition, but from this it could not be concluded that they would fall with the same acceleration in vacuum. Only experiments show this. Moreover, there is nothing in Mach's operational definition of inertial mass ("The mass-ratio of any two bodies is the negative inverse ratio of the mutually induced accelerations of those bodies") which might indicate a connection between inertia and weight (or between m_i and m_g). For this reason Mach's statement (that from his definition of mass he could deduce that mass might be measured by weight) is not warranted. In this regard Newton was on better ground than Mach. According to Newton, it comes *from experience* (of free fall or with pendulums) that we can measure mass by weight.

Mach proposed that the distant matter (such as the fixed stars) establishes a very good inertial system. But he did not explain how this connection between the distant stars and the locally determined inertial frames might arise. He stimulated thinking in the right direction, although he did not supply the key to unlock the mystery.

Another point is that he did not show how the spinning set of stars can generate centrifugal forces. The same can be said of Leibniz, Berkeley and all the others. Mach suggested that nature should behave like this, but he did not propose a specific force law that possessed this property. With Newton's law of gravitation, a spherical shell exerts no forces on internal bodies, whether the shell is at rest or spinning, regardless of the position or motion of the internal bodies. As we will see, this can be implemented with Weber's law for gravitation. To show this quantitative implementation of Mach's ideas is the main goal of this book.

The time was ripe during Mach's life for an implementation of relational mechanics. Physical science was highly developed during the second half of last century. Weber's relational force for electromagnetism appeared in 1846. Mach mentioned this work of Weber in his article *On the fundamental concepts of electrostatics*, delivered in 1883 [84, pp. 107-136, see especially p. 108]. A similar force law was applied to gravitation in the early 1870's, just at the time Mach was publishing his criticisms of Newtonian mechanics and proposing his new formulation. Mach worked with many branches of physics, including mechanics, gravitation, thermodynamics, physiology, acoustics and optics. As regards electromagnetism, his doctoral thesis (1860) was on electrical charge and induction. Other people at that time knew Weber's theory and did not make the connection between Mach's ideas and Weber's work. If someone had had the right insight at that time and connected these two aspects, relational mechanics might have arisen a hundred years before. All the ideas, concepts, force laws and mathematical tools to implement relational mechanics were available during the second half of last century. But it simply did not happen at that time, as history shows. Relational mechanics was not discovered until many years later.

Before entering the new world view of relational mechanics, we will first present Einstein's theories of relativity and the problems it has inflicted on physics. This is the subject of the next chapter.

Chapter 7

Einstein's Theories of Relativity

7.1 Introduction

Albert Einstein (1879-1955) published his special theory of relativity in 1905, while his general theory of relativity was published in 1916. In developing these theories he was greatly influenced by Mach's book *The Science of Mechanics* [85, pp. 282-288]. In the last 80 years physics, and mechanics in particular, have been dominated by Einstein's ideas, since he became famous after 1919 as a result of the solar eclipse expedition, which apparently confirmed his predictions for the bending of light. Newtonian mechanics has since been considered only as an approximation of the "correct" Einsteinian theories.

Here we will argue that Einstein's theories do not implement Machian relational ideas. Moreover, we will show that Relational Mechanics describes the observed phenomena of nature in a better way than Einstein's special and general theories of relativity.

7.2 Einstein's Special Theory of Relativity

Einstein's special theory of relativity is presented in his paper of 1905 entitled "On the electrodynamics of moving bodies" [86, pp. 35-65]. He begins this paper with the following two paragraphs:

> It is known that Maxwell's electrodynamics—as usually understood at the present time - when applied to moving bodies, leads to asymmetries which do not appear to be inherent in the phenomena. Take,

for example, the reciprocal electrodynamic action of a magnet and a conductor. The observable phenomenon here depends only on the relative motion of the conductor and the magnet, whereas the customary view draws a sharp distinction between the two cases in which either the one or the other of these bodies is in motion. For if the magnet is in motion and the conductor at rest, there arises in the neighbourhood of the magnet an electric field with a certain definite energy, producing a current at the places where parts of the conductor are situated. But if the magnet is stationary and the conductor in motion, no electric field arises in the neighbourhood of the magnet. In the conductor, however, we find an electromotive force, to which in itself there is no corresponding energy, but which gives rise - assuming equality of relative motion in the two cases discussed - to electric currents of the same path and intensity as those produced by the electric forces in the former case.

Examples of this sort, together with the unsuccessful attempts to discover any motion of the earth relatively to the "light medium," suggest that the phenomena of electrodynamics as well as of mechanics possess no properties corresponding to the idea of absolute rest. They suggest rather that, as has already been shown to the first order of small quantities, the same laws of electrodynamics and optics will be valid for all frames of reference for which the equations of mechanics hold good. We will raise this conjecture (the purport of which will hereafter be called the "Principle of Relativity") to the status of a postulate, and also introduce another postulate, which is only apparently irreconcilable with the former, namely, that light is always propagated in empty space with a definite velocity c which is independent of the state of motion of the emitting body. These two postulates suffice for the attainment of a simple and consistent theory of electrodynamics of moving bodies based on Maxwell's theory for stationary bodies. The introduction of a "luminiferous ether" will prove to be superfluous inasmuch as the view here to be developed will not require an "absolutely stationary space" provided with special properties, nor assign a velocity-vector to a point of empty space in which electromagnetic processes take place.

Einstein and his followers have introduced many problems into physics with this theory. We discuss a few of them in separate subsections.

7.2.1 Asymmetry in Electromagnetic Induction

The asymmetry of electromagnetic induction noted above does not appear in Maxwell's electromagnetism, contrary to what Einstein wrote. It appears only in Lorentz's formulation of electrodynamics. This asymmetry did not exist for Faraday (1791-1867), who discovered the phenomenon. In 1831 Faraday found that he could induce an electrical current in a secondary circuit if he varied the current in a primary circuit, but that while the current in the primary circuit was constant no induction would be produced [87, see especially Series I, §10]. He also discovered that he could induce a current in the secondary circuit with a constant current in the primary circuit, provided that he moved one or the other relative to the laboratory, so that a relative motion between them would result ([87, see for instance Series I, §18 and 19]). He could also induce a current in the secondary circuit approaching or receding a permanent magnet, or by keeping the magnet at rest relative to the earth and moving the secondary circuit (see for instance §39-43 and 50-54). In order to explain his observations, he arrived at a law of induction [87, §114, p. 281].

According to Faraday, the explanation of the induction when a circuit is moved toward a magnet or *viceversa* is based on the real existence of magnetic lines of force and on their cutting the electrical circuit. Faraday never doubted that these lines of force participated totally in the magnet's translational motion (although he was in doubt as regards rotational motion) [88, p. 155]. To Faraday, if we move a magnet (or current carrying wire) with a constant linear velocity relative to the laboratory, the lines of magnetic field (or lines of force) will also move with the same constant velocity relative to the laboratory, following the motion of the magnet. In the case of rotation he was not so explicit, but since in the experiment described by Einstein there are only translational motions with constant velocities, we will not discuss here the cases of rotation of the magnet.

Maxwell (1831-1879) shared Faraday's views on this matter and did not see any "sharp distinction" in the explanation of Faraday's experiments, whether or not the circuit or magnet moved relative to the laboratory. For instance, in §531 of his *Treatise on Electricity and Magnetism*, he condensed Faraday's experiments into a single law [89, p. 179]. That the lines of force (or lines of magnetic induction, or lines of the magnetic field $\vec{B}$) move relative to the laboratory when the magnet moves relative to the Earth is stated plainly by Maxwell in §541 of his *Treatise*. In Maxwell's view the explanation for the induction in the secondary circuit is always the same, depending only on the relative motion between this secondary circuit and the lines of magnetic field generated by the magnet or primary current-carrying circuit.

Nor does this asymmetry pointed out by Einstein appear in Weber's electrodynamics, although Weber's electrodynamics does not utilize the concept of lines of force or lines of magnetic field $\vec{B}$. Weber's electrodynamics depends only on the relative distances, relative velocities and relative accelerations of the interacting charges ([12, Chapter 3 and Section 5.3]). The concepts of electric and magnetic fields do not need to be introduced in Weber's electrodynamics. Apparently Einstein knew nothing about Weber's electrodynamics, as there is no work by Einstein in which he mentions either Wilhelm Weber's name or Weber's electrodynamics, to the best of our knowledge. Weber's electrodynamics was the main electromagnetic theory in Germany during the third quarter of 19th century. It was also discussed at length in the last Chapter of Maxwell's *Treatise*. It also appears that Einstein never read Maxwell's *Treatise* either, even though it was first published in 1873 and a German translation appeared in 1893 [88, pp. 138-139, note 7].

The phenomenon of induction is always interpreted in the same way in Weber's electrodynamics, whether the observer (or the earth) is at rest relative to the magnet or to the electrical circuit. The only important quantity is the relative velocity between the magnet and the electrical circuit in which the current is induced. The velocity of each of these bodies (magnet or electrical circuit) relative to the observer or to the earth is meaningless in Weber's electrodynamics.

Here we present briefly an analysis of this experiment based on Weber's electrodynamics. The magnet is represented by a circuit 1 in which a current I_1 flows. We want to know the current I_2 which will be induced in a second circuit 2 due to their relative motion. We then consider two rigid circuits 1 and 2 which move relative to the earth with linear velocities $\vec{V_1}$ and $\vec{V_2}$, respectively, without any rotation, as in Figure 7.1.

If there are no batteries or other current sources connected to circuit 2, and if its resistance is R_2, then the induced current which will flow on it due to induction by the first circuit is given by $I_2 = emf_{12}/R_2$, where emf_{12} is the electromotive force induced by the first circuit on the second. Weber's electrodynamics gives the infinitesimal d^2emf_{12} exerted by a neutral current element $I_1 \vec{dl_1}$ (with charges dq_{1+} and $dq_{1-} = -dq_{1+}$) located at $\vec{r_1}$ on another neutral current element $I_2 \vec{dl_2}$ (with charges dq_{2+} and $dq_{2-} = -dq_{2+}$) located at $\vec{r_2}$ as (see [12, Section 5.3] and Figure 7.2):

$$d^2emf_{12} = -\frac{dq_{1+}}{4\pi\varepsilon_o} \frac{\hat{r}_{12} \cdot \vec{dl_2}}{r_{12}^2 c^2} \left\{ 2\vec{V}_{12} \cdot (\vec{v}_{1+d} - \vec{v}_{1-d}) \right.$$

$$\left. - 3(\hat{r}_{12} \cdot \vec{V}_{12})[\hat{r}_{12} \cdot (\vec{v}_{1+d} - \vec{v}_{1-d})] + \vec{r}_{12} \cdot (\vec{a}_{1+} - \vec{a}_{1-}) \right\} .$$

Here r_{12} is the distance between the current elements, $\hat{r}_{12}$ is the unit vector pointing from 2 to 1, $\vec{V}_{12} \equiv \vec{V_1} - \vec{V_2}$, $\vec{v}_{1+d}$ and $\vec{v}_{1-d}$ are the drifting velocities of

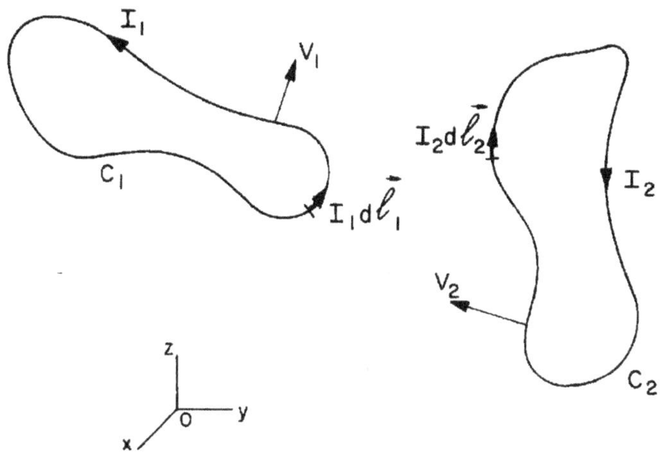

Figure 7.1: Two current carrying circuits moving relative to the earth with velocities V_1 and V_2.

the positive and negative charges of current element 1 (their velocity relative to the wire), while $\vec{a}_{1+}$ and $\vec{a}_{1-}$ are their accelerations relative to the terrestrial frame being considered here.

Integrating this result over the closed circuits C_1 and C_2 yields the usual expression of Faraday and Neumann, namely (see [12], Section 5.3):

$$emf_{12} = -\frac{\mu_o}{4\pi}\frac{d}{dt}\left[I_1 \oint_{C_1}\oint_{C_2}\frac{(\hat{r}_{12}\cdot d\vec{l}_1)(\hat{r}_{12}\cdot d\vec{l}_2)}{r_{12}}\right]$$

$$= -\frac{\mu_o}{4\pi}\frac{d}{dt}\left[I_1 \oint_{C_1}\oint_{C_2}\frac{d\vec{l}_1\cdot d\vec{l}_2}{r_{12}}\right] = -\frac{d}{dt}(I_1 M) \ .$$

Here M is the coefficient of mutual induction.

Supposing I_1 to be a constant in time and rigid circuits which translate as a whole without rotation, with velocities $\vec{V}_1$ and $\vec{V}_2$, this *emf* can be written as:

$$emf_{12} = -I_1 \oint_{C_1}\oint_{C_2} d\vec{l}_1 \cdot d\vec{l}_2 \frac{d}{dt}\frac{1}{r_{12}}$$

$$= I_1(\vec{V}_1 - \vec{V}_2)\cdot \oint_{C_1}\oint_{C_2}\frac{(d\vec{l}_1\cdot d\vec{l}_2)\hat{r}_{12}}{r_{12}^2} \ .$$

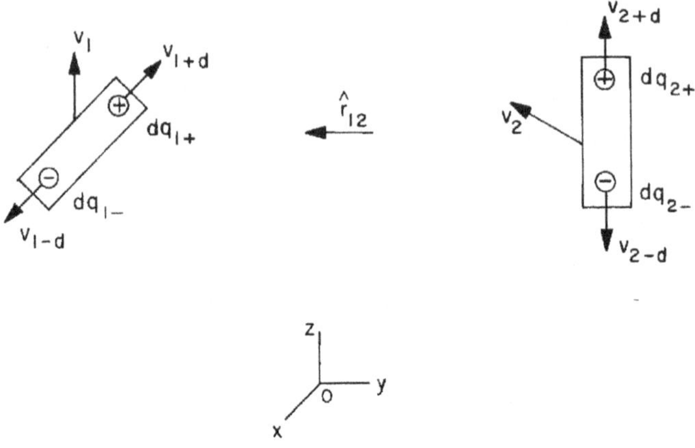

Figure 7.2: Two neutral current elements moving relative to the earth with velocities V_1 and V_2.

This result can also be obtained directly from Weber's electrodynamics with the energy of interaction between the circuits.

What is important to realize is that Weber's expression depends only on the relative velocity between the circuits, $\vec{V}_1 - \vec{V}_2$. This shows that whenever this relative velocity is the same, the induced current will also be the same. For instance, in Einstein's first situation we have the magnet in motion relative to the earth or laboratory and the circuit at rest ($\vec{V}_1 = \vec{V}$ and $\vec{V}_2 = 0$), while in the second situation we have the magnet at rest and the conductor moving in the opposite direction relative to the earth or laboratory ($\vec{V}_1 = 0$ and $\vec{V}_2 = -\vec{V}$). As the relative motion is the same in both cases, $\vec{V}_1 - \vec{V}_2 = \vec{V}$, Weber's law predicts the same induced current, and this is what is observed.

The difference between Weber's prediction and Lorentz's is that we do not need to speak of electric and magnetic fields, so that the Weberian explanation is exactly the same in both cases. There is no "sharp distinction" in the explanation of the induction in both cases.

This distinction appears only in Lorentz's formulation of Maxwell's electrodynamics. Einstein was referring to this formulation when he spoke of the asymmetry in the explanation of the phenomenon ([88, p. 145]). According to Lorentz, when the magnet is in motion with a velocity $\vec{v}_m$ relative to the ether, it generates in the ether not only a magnetic field but also an electric field given by $\vec{E} = \vec{B} \times \vec{v}_m$. This electric field acts in the circuit which is at rest relative to the ether, inducing a current in it. If the magnet is at rest in the ether, it generates only a magnetic field $\vec{B}$ and no electric field, so that when the circuit is moving in the ether with a velocity $\vec{v}_c$, its charges will experience a magnetic force given by $q\vec{v}_c \times \vec{B}$ which will induce a current in the circuit. If $\vec{v}_m = -\vec{v}_c$ then the induced current will be the same. But the origin of this current is completely different in both cases in Lorentz's theory. In the first case, it is due to an electric field, and there is no magnetic force, while in the second case there is no electric field, and the induction is due to a magnetic force. It would appear that Einstein was following the discussion of the phenomenon as presented by Föppl in his book of 1894, which Einstein read between 1896-1900 ([88, pp. 146 and 150-4]).

To Lorentz only velocities relative to the ether were important. By making the ether concept superfluous, Einstein needed to utilize velocities relative to the observer in this analysis. This is the beginning of the introduction in physics of quantities which depend on the observer (or on motion relative to the observer, or frame-dependent quantities). Moreover, by relying on Lorentz's views of Maxwell's electrodynamics, with all the asymmetries built into this formulation, Einstein maintained problems which were to accumulate in the future. All of this might have been avoided if he had opted for the points of view of Faraday, the original viewpoint of Maxwell, or of Weber, or if he had been guided only by the experiments of induction, which do not suggest any asymmetry.

This is one of the strong points in favour of Weber's electrodynamics. There are many other experiments which can be easily explained in this formulation, as is the case of unipolar induction [19].

7.2.2   Postulate of Relativity

Einstein called the postulate of relativity the statement that "the same laws of electrodynamics and optics will be valid for all frames of reference for which the equations of mechanics hold good." He gave the following formal definition for the principle of relativity [86, p. 41]: "The laws by which the states of physical systems undergo changes are not affected, whether these changes of state be referred to the one or the other of two systems of coordinates in uniform translatory motion." This postulate is limited. The reason for this limitation is that in non-inertial frames of reference, Newton's second law of motion in the form of Eq. (1.3) needs to be modified by the introduction of "fictitious forces," as we saw in chapter 3.

Although he called this a postulate of "relativity," this is not the case at all. After all, he retains the Newtonian concept of absolute space disconnected from distant matter. Newton was much more precise and correct when he adopted the words "absolute space and time" to explain his laws of motion. He also knew how to distinguish very clearly the differences which should appear in the phenomena when there was only a relative rotation between local bodies and the fixed stars, or when there was a real absolute rotation of local bodies relative to space (the bucket experiment, the flattenig of the earth, *etc.*)

To Newton, absolute space and the frames of reference not accelerated relative to it form a privileged set of inertial frames, in which the laws of mechanics take their simplest forms. Einstein's postulate of relativity continues to give this set of frames of reference privileged status. For this reason it might be called, more appropriately, as inertial postulate or absolute postulate.

7.2.3 Twin Paradox

We can also see this absolute aspect of Einstein's theory in one of the famous paradoxes which appears in special relativity (but not in Newtonian mechanics nor in the relational mechanics presented here). A detailed discussion of this paradox can be found elsewhere ([90] and [91]).

Two twins A and B are born on the same day on the earth. Later on A travels to a distant place and returns to meet his brother who remained on the earth. According to relativity, the time runs slower for A than for B, so that when they meet again B is older than A. But from the point of view of A, it was B who travelled far away and returned back, so that it should be B who became younger. This is the paradox. To avoid the paradox we might say that they always kept the same age, but this is not what Einstein's theory of relativity predicts. According to this theory, A really becomes younger than B. We can only understand this by saying that while B remained at rest or in uniform rectilinear motion relative to absolute space or to an inertial frame, the same is not true for A, who was in motion and accelerated relative either to absolute space or to an inertial system. Once more, we see that despite the name "relativity," Einstein's theory retains the basic absolute concepts of Newtonian mechanics.

Here we are only discussing the conceptual aspects of Einstein's theory. It is usually stated that this dilation of the proper time of a body in motion has been proved by experiments in which unstable mesons are accelerated and move at high velocities in particle accelerators. In these experiments it is observed that the half-lives (time for radioactive decay) of these accelerated mesons are greater than the half-lives of mesons at rest in the laboratory.

But this is not the only interpretation of these experiments. It can be equally argued that these experiments only show that the half-lives of the unstable mesons depend on their accelerations and high velocities relative to the distant matter in the cosmos, or on the strong electromagnetic fields to which they were subject. Recently Phipps derived this alternative explanation based on relational mechanics [92].

An analogy to this new interpretation is what happens to a common pendulum clock. Suppose two identical pendulum clocks at rest on the earth, marking the same time at sea level and running at the same pace. We then carry one of them to a high mountain, keep it there for several hours, and bring it back to sea level at the location of the other clock. Comparing the two clocks it is observed that the clock which was carried to the top of the mountain is delayed relative to the one which stayed all time at sea level. This is the observational fact. It can be interpreted saying that time ran more slowly for the clock at the top of the mountain. Or it can be interpreted by saying that time ran equally to both clocks, but that the period of oscillation of the pendulum clock depends on the gravitational field of the earth ($T = 2\pi\sqrt{\ell/g}$). As the gravitational field is weaker at the top of the mountain than at sea level, the clock which stayed on the mountain is delayed as compared with the one at sea level. This latter interpretation seems to us more natural and simple (and for this reason more in agreement with the usual procedures of physics) than the other one which involves changes in the fundamental concepts of space and time.

The same reasoning can be applied to the meson experiment. Instead of saying that time runs more slowly to a body in motion, it seems to us more simple to state that the half-lives of the mesons depend on their high velocity relative to the distant material universe. This explanation is not only more suitable to explain the experiment, but also more in agreement with the standard procedures of physics. It also leads to important new suggestions which might be checked experimentally (a possible influence of gravitation on radioactive processes *etc.*)

7.2.4 Constancy of the Velocity of Light

Einstein's second postulate (constancy of the velocity of light) introduces another absolute concept or entity in mechanics, the velocity of light. Here is this postulate [86, p. 38]: "light is always propagated in empty space with a definite velocity c which is independent of the state of motion of the emitting body." On page 41 he gave a formal definition: "Any ray of light moves in the 'stationary' system of co-ordinates with the determined velocity c, whether the ray be emitted by a stationary or by a moving body."

With this postulate it appears that he is advocating the luminiferous ether. After all, the property of something with a constant velocity independent of the motion of the source is characteristic of waves moving in a medium, as is the case of sound moving in air. But soon after the presentation of this postulate he states that "the introduction of a 'luminiferous ether' will prove to be superfluous inasmuch as the view here to be developed will not require an 'absolutely stationary space' provided with special properties." With this statement we can only conclude that for Einstein the velocity of light is constant not only whatever the state of motion of the emitting body, but also whatever the state of motion of the receiving body (detector) and of the observer. This conclusion is confirmed by Einstein's own derivation of this "fact" in another section of his paper. Einstein calls K the stationary inertial system of reference, with coordinates $(x,\ y,\ z,\ t)$, where light propagates at a constant velocity c. The frame of reference k, with coordinates $(\xi,\ \eta,\ \zeta,\ \tau)$, moves relative to K with a constant velocity v along the positive x direction. Here is Einstein's proof [86, p. 46]:

> We now have to prove that any ray of light, measured in the moving system, is propagated with the velocity c, if, as we have assumed, this is the case in the stationary system; for we have not as yet furnished the proof that the principle of the constancy of the velocity of light is compatible with the principle of relativity.
>
> At the time $t = \tau = 0$, when the origin of the co-ordinates is common to the two systems, let a spherical wave be emitted therefrom, and be propagated with the velocity c in system K. If $(x,\ y,\ z)$ be a point just attained by this wave, then
>
> $$x^2 + y^2 + z^2 = c^2 t^2 \ .$$
>
> Transforming this equation with the aid of our equations of transformation we obtain after a simple calculation
>
> $$\xi^2 + \eta^2 + \zeta^2 = c^2 \tau^2 \ .$$
>
> The wave under consideration is therefore no less a spherical wave with velocity of propagation c when viewed in the moving system. This shows that our two fundamental principles are compatible.

To us this is the main problem with Einstein's theory. The reason is the following: In everyday life we know two kinds of phenomena in physics. The first kind is ballistic phenomena. Suppose that a cannon at rest on the earth's surface shoots cannon balls with a certain initial velocity v_b relative to the earth, neglecting the effects of air resistance. In this analysis we will also neglect the effect of gravity, which would bend the trajectories of the bullets in parabolic orbits. If the cannon moves with a velocity v_c relative to the earth and shoots a cannon ball, the velocity of the bullet relative to the earth will be $v_b + v_c$, while the velocity of the bullet relative to the cannon will still be v_b, once more neglecting the effects of air resistance. This is typical of ballistic effects. Now suppose we have a man holding several identical guns at rest relative to the earth, each one of them pointing in one direction. Shooting all of them at the same time will produce a spherical surface of bullets moving with velocity v_b relative to him and to the earth, as in Figure 7.3.

Now if this man moves with a velocity v_c relative to the earth's surface and shoots all the guns he is holding at the same time, he will still see a spherical surface of bullets moving away from him with a velocity v_b, neglecting the effects of air resistance. Let us represent the earth's frame by O with coordinates (x, y) and the person's frame (moving relative to the earth) by O' with coordinates (x', y'). In this case the equation describing the spherical surface of bullets centered on the moving man is given by $x'^2 + y'^2 = (v_b t)^2$, where t is the time since the shooting of the guns. With the man moving to the right in the x direction with velocity v_c we have: $x' = x - v_c t$ and $y' = y$, such that the equation of the surface of bullets relative to the earth at the place where the man shot the guns is given by $(x - v_c t)^2 + y^2 = (v_b t)^2$. For an observer who stayed at rest relative to the earth the surface of the bullets will only be centered on him at the initial instant, as in Figure 7.4, where all velocities are relative to the earth's surface. The form of the equation changed, and is no longer given by $x^2 + y^2 = (v_b t)^2$, although this was the form of the equation in the moving frame. We can see that Einstein's conclusion (that the form of the wave equation is invariable) is not valid for ballistic effects. Moreover, in these ballistic effects the velocity of the bullet depends directly on the velocity of the source.

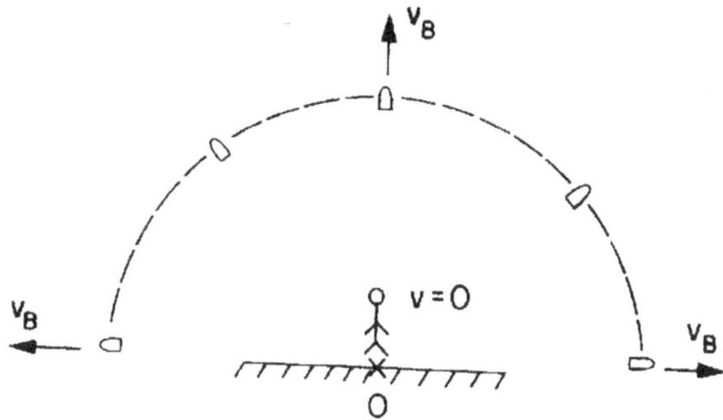

Figure 7.3: Man shooting bullets while at rest relative to the earth.

The other kind of phenomenon known in physics depends directly on the medium. The simplest example is that of sound. Suppose we have a train at rest relative to the earth emitting a sound which moves relative to the earth with the velocity v_s, assuming the air to be at rest relative to the earth, as in Figure 7.5.

If the train now moves with a velocity v_t relative to the earth and emits a sound, the sound will still move with the velocity v_s relative to the earth. But

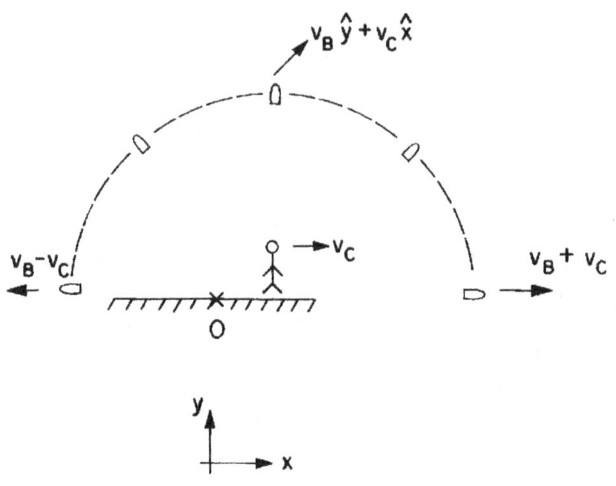

Figure 7.4: Man shooting bullets while moving relative to the earth.

now the velocity of the sound relative to the train will be $v_s - v_t$ in the forward direction and $v_s + v_t$ in the backward direction, assuming once more that the air is at rest relative to the earth and that $v_s > v_t$.

In the case of sound, the form of the sound wave relative to the earth is always spherical from the point of emission, whether the train is at rest or moving relative to the earth: $x^2 + y^2 = (v_s t)^2$. Relative to the train moving in the x direction with a velocity v_t (frame O') the equation of the sound wave takes the form $(x' + v_t t)^2 + y'^2 = (v_s t)^2$ and not $x'^2 + y'^2 = (v_s t)^2$. Now the spherical surface is no longer centered on the moving frame (the train), shown in Figure 7.6, where all velocities are relative to the earth's surface. Einstein's conclusion that the form of the wave equation is invariable is not valid in the case of sound either. Despite this fact, the velocity of sound relative to the earth is independent of the state of motion of the source relative to the earth.

In the ballistic case the velocity of the bullets is constant relative to the source at the moment of emission, even when the source is moving relative to the earth. On the other hand, in the case of the whistling train the velocity of sound is constant relative to the air and does not depend on the velocity of the source. The form of the equation describing the wave front changes for different moving frames in both cases.

Moreover, the velocity of the sound and of the bullets depends on the velocity of the observer or detector. Let us consider an observer or detector O moving

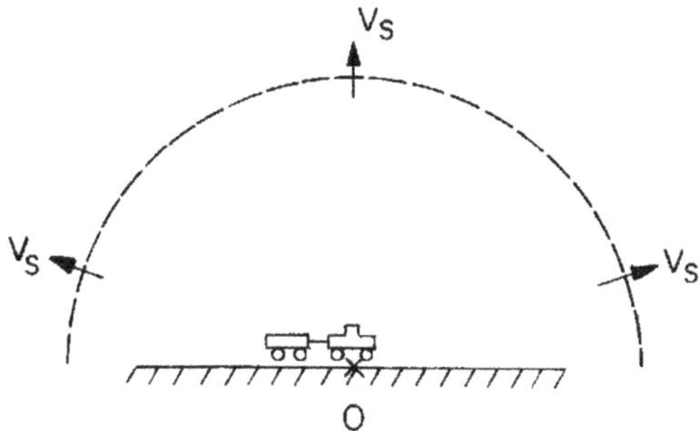

Figure 7.5: Train blowing whistle while at rest relative to the earth and air.

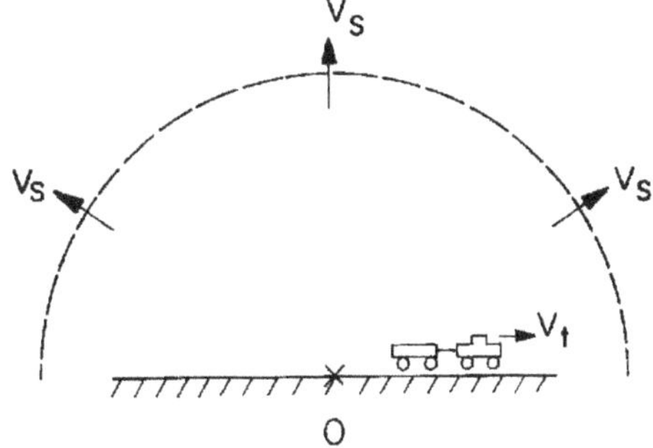

Figure 7.6: Train blowing whistle while moving relative to the earth and air.

with a velocity $\vec{v}_o = -v_o\hat{x}$ relative to the earth. In the ballistic case he will find a velocity $v_b + v_o$ in the first case (cannon at rest relative to the earth shooting a gun with velocity $+v_b\hat{x}$ relative to the earth) and $v_b + v_c + v_o$ in the second case (cannon moving with velocity $+v_c\hat{x}$ relative to the earth), shown in Figures 7.7 and 7.8, where all velocities are relative to the earth.

As regards sound, in both cases (train at rest or moving relative to the earth) the observer or detector will find the velocity of sound given by $v_s + v_o$, irrespective of the train's velocity relative to the earth or the air, as in Figures 7.9 and 7.10, where all velocities are relative to the earth and air. Once more we are supposing the air at rest relative to the earth.

Let us give just another example of reasoning which shows that light velocity must depend on the velocity of the observer or detector. Consider first a man

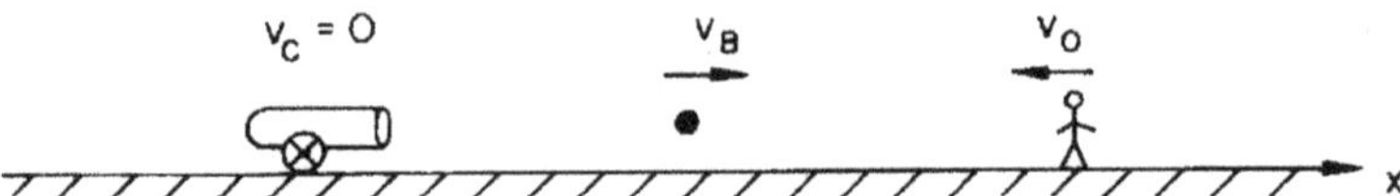

Figure 7.7: Observer moving relative to the earth while the cannon is at rest.

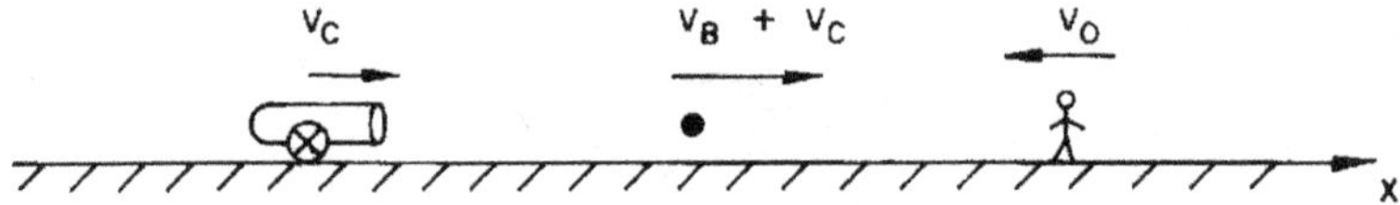

Figure 7.8: Observer and cannon moving relative to the earth.

with a gun at rest in the laboratory. We will neglect the effect of air friction and the deflection of the bullet due to the gravitational attraction of the earth. Let us suppose that the man shoots a bullet with a velocity of 30 m/s. It will take one second for the bullet to cross a 30 meter long room. Now let us suppose there is a person at each end of the room. If both of them shoot their guns toward each other at the same time, in half a second each bullet will move 15 meters, and the two will meet at the center of the room. The velocity of one bullet relative to the other is obviously 60 m/s, as they moved the same 30 meters (15 meters each) but only in half a second. Alternatively, if the two persons shot the guns at the same time but in the same direction, the two bullets will never meet, keeping the same distance from each other, no matter

Figure 7.9: Observer moving relative to the earth and air, while the train is at rest.

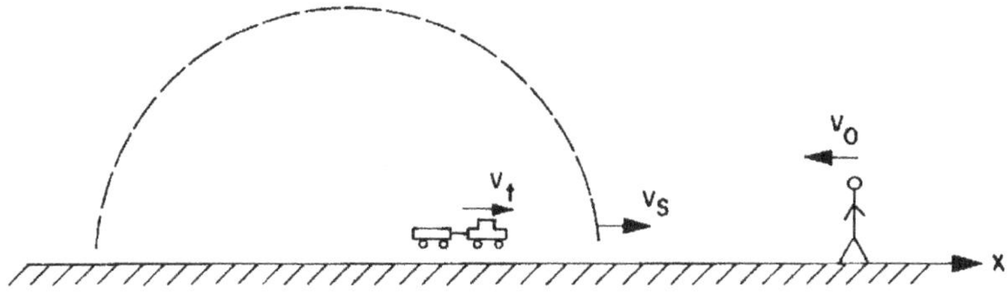

Figure 7.10: Observer and train moving relative to the earth and air.

how long we wait. This means that in this second situation the velocity of one bullet relative to the other is obviously zero, as their distance remains constant as time passes. In order to see that the same must be true for light, we only need to replace the words "gun" with "light source (like a lantern)," "bullet" with "photon (or wave front)" and "30 m/s" by "3×10^8 m/s." In the first case the velocity of one photon relative to the other must be 6×10^8 m/s, while in the second case it must be zero. After all, if two photons (or wave fronts) move in the same direction with the same velocity, the velocity of one photon relative to the other must be zero, by the definition of velocity (change of distance by time interval). It cannot be 3×10^8 m/s as Einstein said.

As we see here, the velocity of the cannon balls or of the sound always depends on the velocity of the observer or of the detector. But what Einstein concluded is that light is a completely different entity, such that its velocity in vacuum never depends on the velocity of the observer. However, light is a physical entity which carries momentum and energy, is affected by the medium in which it propagates (reflection, refraction, difraction, Faraday rotation of the plane of polarization *etc.*), it acts on bodies (heating them, causing chemical reactions, ionizing atoms *etc.*) In this sense it is not a special entity. As such it has certain similarities with both bullets and waves. Acceptance of the conclusion that light velocity in vacuum is a constant for all inertial observers, irrespective of their motion relative to the source, has created a host of problems and paradoxes in the last 90 years.

To prove that the velocity of light does not depend on the motion of the observer or detector, it would be necessary to perform experiments in the laboratory in which the detector was moving at high velocities (close to c) relative to the earth, while the source of light was at rest in the earth. To our knowledge this kind of experiment has never been performed.

Wesley, Tolchelnikova-Murri, Hayden, Monti and several other authors have presented strong and convincing arguments that the methods employed by Roemer and by Bradley to obtain the velocity of light prove that the measured value of this velocity depends on the velocity of the observer relative to the source: Wesley [93, Sections 2.2: Roemer's measurement of the velocity of light and 2.4: Bradley aberration to measure the velocity of light], Tolchelnikova-Murri [94], Hayden [95] and Monti [96]. Roemer's fundamental work can be found in the original French ([97] and in [98, pp. 151-154]) and in English translation ([99] and [97]). Bradley's work can be found in English: Bradley [100] and [101], and Sarton [102].

As we have seen, Einstein maintained the Newtonian concept of absolute space (or of preferred inertial frames of reference) independent of distant matter and introduced another absolute quantity in the theory, light velocity in vacuum. The works of Wesley and of Monti, on the other hand, show that light velocity is a function of the state of motion of the observer.

7.2.5 Velocity in Lorentz's Force

Another problem created by Einstein was his interpretation of the velocity in the magnetic force $q\vec{v} \times \vec{B}$. We discussed this problem previously ([103] and [12, Appendix A: The Origins and Meanings of the Magnetic Force $\vec{F} = q\vec{v} \times \vec{B}$], where all the relevant references can be found).

In Lorentz's force, relative to what object, body or entity is to be understood the velocity $\vec{v}$ of the charge q? Some options: relative to the macroscopic source of the magnetic field (a magnet or current carrying wire), relative to the magnetic field itself, relative to an inertial frame of reference, relative to an arbitrary frame of reference not necessarily inertial, relative to the laboratory or to the earth, relative to the average motion of the microscopic charges (usually electrons) giving rise to the magnetic field, relative to the detector of the magnetic field, *etc.* Here we present the historical interpretations of this velocity.

Maxwell died in 1879. In 1881 J. J. Thomson (1856-1940) for the first time in physics formulated theoretically the magnetic force as $q\vec{v} \times \vec{B}/2$ [104, pp. 306-310]. The velocity $\vec{v}$ in this theory was the velocity of the charge q relative to the medium through which it was moving, a medium whose magnetic permeability was μ. For Thomson this velocity of q was not relative to the luminiferous ether, nor relative to the magnet which generated $\vec{B}$, nor the velocity of q relative to the observer. He called this $\vec{v}$ the *actual velocity* of the particle of charge q. On page 248 of his article he stated: "It must be remarked that what we have for convenience called the actual velocity of the particle is, in fact, the velocity of the particle relative to the medium through which it is moving" ..., "medium whose magnetic permeability is μ" [105]. In 1889, O. Heaviside (1850-1925) deduced theoretically $q\vec{v} \times \vec{B}$ (twice the value obtained by Thomson). He accepted Thomson's interpretation for the meaning of $\vec{v}$, as can be seen from the title of his paper: "On the electromagnetic effects due to the motion of electrification through a dielectric" [106]. This title shows that for him this $\vec{v}$ was the velocity of the charge q relative to the dielectric material through which it was moving. H. A. Lorentz (1853-1928) presented his famous force law $\vec{F} = q\vec{E} + q\vec{v} \times \vec{B}$ in 1895. To our knowledge he never performed a single experiment to arrive at his expression. In order to show how he arrived at it, we present the discussion in Lorentz's famous book *The Theory of Electrons*. Passages in square brackets are our words and the modern rendering of some of his formulas (for instance, [**a** · **b**] is nowadays usually represented by $\vec{a} \times \vec{b}$). He utilized the cgs system of units. What he called "electron" represented a generic electrical particle (the charge we call nowadays "electron," with a charge $q = -1.6 \times 10^{-19}$ C and mass $m = 9.1 \times 10^{-31}$ kg, was only discovered in 1897). We emphasized some important words. See Lorentz [107, pp. 14-15]:

> However this may be, we must certainly speak of such a thing as the force acting on a charge, or on an electron, on charged matter, whichever appelation you prefer. Now, in accordance with the general principles of Maxwell's theory, *we shall consider this force as caused by the state of the ether, and even, since this medium pervades the electrons, as exerted by the ether on all internal points of these particles where there is a charge.* If we divide the whole electron into elements of volume, there will be a force acting on each element and determined by the state of the ether existing within it. We shall suppose that this force is proportional to the charge of the element, so that we only want to know the force acting per unit charge. This is what we can now properly call *the electric force.* We shall represent it by **f**. The formula by which it is determined, and which is the one we still have to add to (17)-(20) [Maxwell's equation's], is as follows:

$$\mathbf{f} = \mathbf{d} + \frac{1}{c}[\mathbf{v} \cdot \mathbf{h}]. \qquad \left[\vec{f} = d' + \frac{\vec{v} \times \vec{h}}{c} \right]. \qquad (23)$$

> Like our former equations, it is got by generalizing the results of electromagnetic experiments. The first term represents the force acting on an electron in an electrostatic field; indeed, in this case, the force per unit charge must be wholly determined by the dielectric displacement. On the other hand, the part of the force expressed by the second term may be derived from the law according to which an element of a wire carrying a current is acted on by a magnetic field

with a force perpendicular to itself and the lines of force, an action, which in our units may be represented in vector notation by

$$\mathbf{F} = \frac{s}{c}[\mathbf{i} \cdot \mathbf{h}], \qquad \left[\vec{F} = \frac{i\,d\vec{l} \times \vec{h}}{c} \right],$$

where $\mathbf{i}$ is the intensity of the current considered as a vector, and s the length of the element. According to the theory of electrons, $\mathbf{F}$ is made up of all the forces with which the field $\mathbf{h}$ acts on the separate electrons moving in the wire. Now, simplifying the question by the assumption of only one kind of moving electrons with equal charges e and a common velocity $\mathbf{v}$, we may write

$$s\mathbf{i} = Ne\mathbf{v},$$

if N is the whole number of these particles in the element s. Hence

$$\mathbf{F} = \frac{Ne}{c}[\mathbf{v} \cdot \mathbf{h}], \qquad \left[\vec{F} = \frac{Ne\vec{v} \times \vec{h}}{c} \right],$$

so that, dividing by Ne, we find for the force per unit charge

$$\frac{1}{c}[\mathbf{v} \cdot \mathbf{h}]. \qquad \left[\frac{\vec{v} \times \vec{h}}{c} \right].$$

As an interesting and simple application of this result, I may mention the explanation it affords of the induction current that is produced in a wire moving across the magnetic lines of force. The two kinds of electrons having the velocity $\mathbf{v}$ of the wire, are in this case driven in opposite directions by forces which are determined by our formula.

9. After having been led in one particular case to the existence of the force $\mathbf{d}$, and in another to that of the force $\frac{1}{c}[\mathbf{v} \cdot \mathbf{h}]$, we now combine the two in the way shown in the equation (23), going beyond the direct result of experiments by the assumption that in general the two forces exist at the same time. If, for example, an electron were moving in a space traversed by Hertzian waves, we could calculate the action of the field on it by means of the values of $\mathbf{d}$ and $\mathbf{h}$, such as they are at the point of the field occupied by the particle.

It is difficult to disagree with O'Rahilly, when he notes that this proof of the formula is extremely unsatisfactory, adding that [108, p. 561]: "There are two overwhelming objections to this alleged generalization. (1) The two 'particular cases' here 'combined' are quite incompatible. In the one case we have charges at rest, in the other the charges are moving; they cannot be both stationary and moving. (2) Experiments with a 'wire carrying a current' have to do with *neutral* currents, yet the derivation contradicts this neutrality."

We can also mention that in his generalization Lorentz did not consider the possibility that the electromagnetic force might depend on the acceleration of the test body, nor on the square of the velocity of the test body. These two terms appear in Weber's force law but not in Lorentz's force.

As we can see from the above quotation ("... force as caused by the state of the ether, and even, since this medium pervades the electrons, as exerted by the ether ..."), for Lorentz it was originally the velocity of the charge relative to the ether and not, for instance, relative to the observer or frame of reference. In Lorentz's theory the ether was in a state of absolute rest relative to the frame of fixed stars [85, p. 111]. A conclusive proof of this interpretation can be found in another work of Lorentz, *Lectures on Theoretical Physics* [109, Vol. 3, p. 306] and [108, Vol. 2, p. 566]. In this work Lorentz says: "Imagine an electric current flowing in a closed circuit without resistance. Will this current act upon a particle carrying a charge e which is placed in its neighbourhood? (...) The answer to this question was, of course, that the current did not act upon the particle. (...) Suppose, however, that both share in some motion, *e. g.* the earth's motion. What then? To begin with, the charged particle will move with a certain velocity through the magnetic field of the current and it will thus be acted upon by some force. (...)" In this last case there is no motion between the charge and the current carrying circuit, nor between the charge and the earth or laboratory, nor even between the charge and the observer (who is supposedly at rest in the laboratory). But to Lorentz, even in this case there will be a magnetic force acting on the charge. He could only consider this because he supposed the $\vec{v}$ to be the velocity of the charge relative to the ether or to the fixed stars. As the fixed stars did not cause any net force on q, all that remains is the force exerted by the ether.

Einstein changed all this with his paper of 1905 on the special theory of relativity. What Einstein proposed in this paper was that the velocity $\vec{v}$ in Lorentz's force should be interpreted as the velocity of the charge relative to the observer (and not relative to the dielectric, as maintained by Thomson and Heaviside, nor relative to the ether, as maintanined by Lorentz). Eintein obtains Lorentz's transformations for the coordinates and for time (transformations which relate the magnitudes in one inertial frame to another inertial frame moving relative to the first with a constant velocity). He obtains these transformations also for the electric and magnetic fields. Einstein applies them in Lorentz's force $\vec{F} = q\vec{E} + q\vec{v} \times \vec{B}$. He then begins to utilize the velocity $\vec{v}$ as the velocity of the charge q relative to the observer. For instance, on page 54 he gives the difference between the old paradigm of electromagnetism and the new one based on his theory of relativity (passages between square brackets are our words):

> Consequently the first three equations above [for the transformation of the electric and magnetic field components in two different inertial systems which move relative to one another] allow themselves to be clothed in words in the following ways:
>
> 1. If a unit electric point charge is in motion in an electromotive field, there acts upon it, in addition to the electric force, an "electromotive force" which, if we neglect the terms multiplied by the second and higher powers of v/c, is equal to the vector-product of the velocity of the charge and the magnetic force $[\vec{B}]$, divided by the velocity of light. (Old manner of expression.)
>
> 2. If a unit electric point charge is in motion in an electromagnetic field, the force acting upon it is equal to the electric force which is present at the locality of the charge, and which we ascertain by transformation of the field to a system of coordinates at rest relatively to the electrical charge. (New manner of expression.)

In other words, according to Einstein we would have $\vec{F} = q\vec{E} + q\vec{v} \times \vec{B}$ in the first case, while in the second case we only have $\vec{F} = q\vec{E}'$, because now $\vec{v}'$ (the velocity of the charge relative to the new system of reference at rest relative to it) is zero, so that $q\vec{v}' \times \vec{B}' = 0$. Einstein is introducing frame-dependent forces here, *i.e.*, forces whose values depend on the motion between the test body and the observer. The introduction of physical forces which depend on the state of motion of the observer has created many problems for the explanation of several simple phenomena of nature. Unfortunately it has been part of theoretical physics ever since that time. No experiment has suggested or forced this new interpretation. This whole interpretation arose from Einstein's mind. The usual expression for the magnetic force might have been maintained, interpreting $\vec{v}$ as the velocity of the test charge relative to the magnet or current-carrying wire, without any contradictions with experimental data.

7.2.6 Michelson-Morley Experiment

Another problem created by Einstein is due to his interpretation of the Michelson-Morley experiment. This famous experiment sought an interference pattern of two light beams which was thought to depend on the motion of the earth relative to the ether. No effect was found with the predicted order of magnitude (experiment with a precision of first order in v/c performed by Michelson in 1881 and of second order performed by Michelson and Morley in 1887, where v was the supposed velocity of the earth relative to the ether, taken in practice as the velocity of the earth relative to the frame of fixed stars).

The most straightforward interpretation of this experiment is that there is no ether. Only the relative motion between light, the mirrors, the charges in them and the earth are important, no matter what the velocity of any of these bodies relative to the ether or to absolute space. In this regard the results obtained by Michelson and Morley agree completely with Weber's electrodynamics, as in this theory the ether plays no role.

As noted earlier, Lorentz (and Fitzgerald) believed in the ether. To reconcile the null result of the experiment with the idea of an ether which is at rest relative to the set of fixed stars, and which was not dragged by the earth, they needed to introduce the idea of length contraction of rigid bodies moving through the assumed ether. This was strange and *ad hoc*, but worked.

Let us see what Lorentz has to say in his text of 1895 [110], our emphasis:

Michelson's Interference Experiment

1. As Maxwell first remarked and as follows from a very simple calculation, the time required by a ray of light to travel from a point A to a point B and back to A must vary when the two points together undergo a displacement without carrying the ether with them. The difference is, certainly, a magnitude of second order; but it is sufficiently great to be detected by a sensitive interference method.

(...)

If we assume the arm which lies in the direction of the Earth's motion to be shorter than the other by $Lv^2/2c^2$, and, at the same time, that the translation has the influence which Fresnel's theory allows it, then the result of the Michelson experiment is explained completely.

Thus one would have to imagine that the motion of a solid body (such as a brass rod or the stone disc employed in the later experiments) *through the resting ether* exerts upon the dimensions of that body an influence which varies according to the orientation of the body with respect to the direction of motion. (...)

Einstein, however, stated that "the introduction of the 'luminiferous ether' will prove to be superfluous." If this is the case, then he should have discarded length contraction of rods and rigid bodies. After all, this idea of length contraction was only introduced to reconcile the null result of the Michelson-Morley experiment with the ether concept. If there is no ether, we should not expect any change in the interference fringes (and no displacement was found with the expected value). But in this case it makes no sense to introduce or to suppose a length contraction of bodies. Making the ether superfluous would require making length contraction superfluous as well. This was clearly pointed out by O'Rahilly in his book, *Electromagnetic Theory - A Critical Examination of Fundamentals*, Vol. 1, Chap. VIII, Sect. 1, p. 259 [108]. As we know, this logical course was not followed by Einstein. He retained the length contraction although he had discarded the ether! With this, another source of confusions and paradoxes was brought into physics.

There are several other problems with Einstein's special theory of relativity: the difficulty in explaining the Sagnac and Michelson-Gale experiments ([93], [95] and [96]); observations of Doppler effects for Venus seem to contradict special relativity ([111]); superluminal solutions of Maxwell's equations challenge the principle of relativity ([112]) *etc.* We will not go into further detail here.

After discussing some aspects of Eintein's special theory of relativity, we analyse his general theory in the next section.

7.3   Einstein's General Theory of Relativity

Einstein's general theory of relativity is presented in his 1916 work called "The foundation of the general theory of relativity" [113]. We present several problems with this theory, as we have done with his special relativity.

7.3.1   Relational Quantities

Einstein begins his article with the following paragraphs:

> The special theory of relativity is based on the following postulate, which is also satisfied by the mechanics of Galileo and Newton.

> If a system of co-ordinates K is chosen so that, in relation to it, physical laws hold good in their simplest form, the *same* laws also hold good in relation to any other system of co-ordinates K' moving in uniform translation relatively to K. This postulate we call the "special principle of relativity." The word "special" is meant to intimate that the principle is restricted to the case when K' has a motion of uniform translation relatively to K, but that the equivalence of K' and K does not extend to the case of nonuniform motion of K' relatively to K.

In the general theory of relativity Einstein sought to generalize his special theory in such a way that "the laws of physics must be of such a nature that they apply to systems of reference in any kind of motion" (his words), and not only for inertial frames [113, p. 113]. Did he succeed in doing so? There are reasons for doubt. One of the reasons for this failure has been the path he chose to follow in order to implement his ideas.

According to Barbour [4, p. 6]:

> Einstein himself commented[1] that the simplest way of realizing the aim of the theory of relativity would appear to be to formulate the laws of motion directly and *ab initio* in terms of relative distances and relative velocities—nothing else should appear in the theory. He gave as the reason for *not* choosing this route its impracticability. In his view the history of science had demonstrated the practical impossibility of dispensing with coordinate systems.

Here is Barbour's translation of the relevant section of Einstein's paper [69, p. 186]:

> We want to distinguish more clearly between quantities that belong to a physical system as such (are independent of the choice of the coordinate system) and quantitities that depend on the coordinate system. One's initial reaction would be to require that physics should introduce in its laws only the quantities of the first kind. However, it has been found that this approach cannot be realized in practice, as the development of classical mechanics has already clearly shown. One could, for example, think—and this was actually done—of introducing in the laws of classical mechanics only the distances of material points from each other instead of coordinates; a priori one could expect that in this manner the aim of the theory of relativity should be most readily achieved. However, the scientific development has not confirmed this conjecture. It cannot dispense with coordinate systems and must therefore make use in the coordinates of quantities that cannot be regarded as the results of definable measurements.

As we will see in this book, it is possible to follow this route successfully with a Weber-type law for gravitation. Einstein was mistaken when he asserted that this route was impractical. Weber introduced his relational force in 1846, 70 years prior to this statement by Einstein. In this book we show that with a Weber-type law applied to gravitation (as suggested by several authors since the 1870's), we can implement quantitatively all of Mach's ideas. We show that by spinning a spherical shell or the distant universe, centrifugal and Coriolis's forces spring into action; we implement a mechanics without absolute space

---

[1]A. Einstein, *Naturwissenschaften*, 6-er Jahrgang, No. 48, 697 (1918) (passage on p. 699).

and time; and also without frame-dependent forces; the inertial frames become directly related or determined by the distant material universe; dynamics becomes equivalent to kinematics; and even the kinetic energy can be shown to be an energy of interaction like any other potential energy.

7.3.2 Invariance in the Form of the Equations

In another part of the paper Einstein explained what he meant by the statement that "the laws of physics must be of such a nature that they apply to systems of reference in any kind of motion." He clarified his thoughts by stating that [113, p. 117]: "The general laws of nature are to be expressed by equations which hold good for all systems of co-ordinates, that is, are co-variant with respect to any substitutions whatever (generally co-variant)." The term covariant had been introduced by Minkowski in 1907-08. He referred to the *identity or equality in the form of the equations* in different inertial frames as "covariance" [88, pp. 14, 240-1 and 288]. Thus, by laws of the same *nature* Einstein meant laws of the same *form*. But this requirement is known to be false when we are dealing with non-inertial frames of reference. For instance, in an inertial frame of reference O we write Newton's second law of motion in the form $\vec{F} = m_i \vec{a}$, while in a non-inertial frame of reference O' which rotates relative to the first one with a constant angular velocity $\vec{\omega}$, this law takes the form: $\vec{F} = m_i(\vec{a}' + \vec{\omega} \times (\vec{\omega} \times \vec{r}') + 2\vec{\omega} \times \vec{v}')$. This works perfectly well, as we saw in chapter 3. This means that Einstein's statement that the laws of physics should have the same form in all frames of reference can only cause confusion and ambiguities. We need to change many concepts of space, time, measurements *etc.* in order for this theory to correctly predict the facts in different accelerated frames of reference. It would be much simpler, more coherent and in agreement with the previous knowledge of the laws of physics to require that each two-body force have the same numerical value (although not necessarily the same form) in all frames of reference. Even Newton's inertial forces have this property. For instance, the value $m_i \vec{a}$ in the inertial system O is exactly equal in magnitude and direction to the value $m_i(\vec{a}' + \vec{\omega} \times (\vec{\omega} \times \vec{r}') + 2\vec{\omega} \times \vec{v}')$ in the non-inertial system O', although the form is completely different in both cases. This is what is implemented in relational mechanics.

7.3.3 Implementation of Mach's Ideas

There are many other problems with Einstein's general theory of relativity. In particular, although he tried to implement Mach's principle with this theory, he did not succeed. In a book originally published in 1922, *The Meaning of Relativity*, Einstein presented three consequences which ought to be expected in any theory implementing Mach's ideas [114, pp. 95-96]:

What is to be expected along the line of Mach's thought?

1. The inertia of a body must increase when ponderable masses are pilled up in its neighbourhood.

2. A body must experience an accelerating force when neighbourring masses are accelerated, and, in fact, the force must be in the same direction as that acceleration.

3. A rotating hollow body must generate inside of itself a 'Coriolis field', which deflects moving bodies in the sense of the rotation, and a radial centrifugal field as well.

We shall now show that these three effects, which are to be expected in accordance with Mach's ideas, are actually present according to our theory, although their magnitude is so small that confirmation of them by laboratory experiments is not to be thought of.

According to Einstein, a fourth consequence which should appear in any theory incorporating Mach's principle was: "4. A body in an otherwise empty universe should have no inertia" [75]. Related to this is the following statement: "4'. *All* the inertia of any body should come from its interaction with other masses in the universe" [75]. A statement of Einstein similar to these two is [114, p. 98]: "Although all of these effects are inaccessible to experiment, because k is so small, nevertheless they certainly exist according to the general theory of relativity. We must see in them a strong support for Mach's ideas as to the relativity of all inertial actions. If we think these ideas consistently through to the end we must expect the *whole* inertia, that is, the *whole* $g_{\mu\nu}$-field, to be determined by the matter of the universe, and not mainly by the boundary conditions at infinity." Another statement of Einstein along the same direction can be found in [115, see especially p. 180]: "In a consistent theory of relativity there can be no inertia *relatively to "space,"* but only an inertia of masses *relatively to one another.* If, therefore, I have a mass at a sufficient distance from all other masses in the universe, its inertia must fall to zero."

Although Einstein at first thought that these consequences did follow from his general theory of relativity, he soon realized this was not the case. For an analysis of this we refer the readers to references in the bibliography (Sciama [70], Reinhardt [75], Raine [77] and Pais [85, pp. 282-8]).

The first consequence does not appear in general relativity. There are no observable effects in a laboratory from a spherically symmetric agglomeration of matter at rest around it. This means that the inertia of a body does not increase in general relativity with the agglomeration of masses in its neighbourhood. Einstein initially arrived at the wrong conclusion that general relativity predicted this effect based on a misinterpretation of a calculation performed in a special coordinate system, as was pointed out by Brans in 1962 (Brans [116], Reinhardt [75] and Pfister [117]).

The second consequence does occur in general relativity, but its interpretation is not unique ([75]).

The third consequence also appears in general relativity, as was first found by Thirring (1888-1976) in 1918 and 1921 ([118] and [119]). There is a translation to English of these basic papers of Thirring and of another one of Lense and Thirring in Mashhoon, Hehl and Theiss [120]. However, the terms obtained by Thirring based on general relativity are not exactly as they should be. Working in the weak field approximation he showed that a spherical shell of mass M, radius R, spinning with a constant angular velocity $\vec{\omega}$ relative to a frame of reference O exerts a force on an internal test particle of mass m, located at $\vec{r}$ relative to the center of the shell, moving with velocity $\vec{v}$ and acceleration $\vec{a}$, given by (Thirring [119], Peixoto and Rosa [121] and Pfister [117]):

$$\vec{F} = -\frac{4GM}{15Rc^2} \left[m\vec{\omega} \times (\vec{\omega} \times \vec{r}) + 10m\vec{v} \times \vec{\omega} + 2m(\vec{\omega} \cdot \vec{r})\vec{\omega} \right] \ . \tag{7.1}$$

The common coefficient $4GM/15Rc^2$ has no dimensions. We will call this expression Thirring's force.

There is an axial term proportional to $(\vec{\omega} \cdot \vec{r})\vec{\omega}$ in this equation which does not have a corresponding term in Newtonian theory. In other words, there is no "fictitious force" which behaves like this.

Moreover, Einstein wanted to obtain the classical centrifugal and Coriolis forces after integrating this result for the whole universe. Considering that the integration of $4GM/15Rc^2$ over the whole universe yields exactly one, we can see that Thirring's force yields the correct centrifugal force. On the other hand, it will yield simultaneously a term 5 times larger than the classical Coriolis force $2m\vec{\omega} \times \vec{v}$. After all, the classical fictitious force $\vec{F}_f$ is given by Eq. (3.17), namely:

$$\vec{F}_f = -m\vec{\omega} \times (\vec{\omega} \times \vec{r}) - 2m\vec{\omega} \times \vec{v}$$

$$- m\frac{d\vec{\omega}}{dt} \times \vec{r} - m\vec{a}_{o'o} \ .$$

This shows that Einstein's general theory of relativity does not succeed in deriving the centrifugal and Coriolis forces simultaneously. Later developments based on general relativity by Bass, Pirani, Brill, Cohen and many others did not succeed as well. They could not derive these two terms simultaneously with the correct coefficients, as they are known to exist in non-inertial frames of reference in Newtonian theory. Moreover, they could not eliminate the spurious axial term at the same time. For discussions and references, see Bass and Pirani

[122], Brill and Cohen [123], Cohen and Brill [124], Reinhardt [75], Peixoto and Rosa [121] and Pfister [117]. This shows that we cannot derive the correct Newtonian results in non-inertial frames of reference with Einstein's theory of relativity.

Putting $\vec{\omega} = 0$ in Thirring's force, Eq. (7.1), once more demonstrates that a stationary and not spinning spherical shell does not exert any force on an internal test particle according to general relativity, no matter what the position, velocity and acceleration of the particle relative to the shell. Consequently, the inertia of a body does not increase when we pile up symmetrically stationary ponderable mass in its neighbourhood, contrary to Einstein's wishes.

It should be observed here that Einstein arrived at the third consequence (that a spinning shell should generate centrifugal forces on bodies in its interior) influenced by Mach's ideas. As we saw previously in section 5.1, Clarke concluded two hundred years before that Leibniz's ideas would lead to the same consequence, but upsidedown. In other words, annihilating the fixed stars (many spherical shells) which are spinning around the solar system should annihilate the centrifugal forces. This shows how similar the ideas of Leibniz and Mach actually are.

The fourth consequence does not arise in general relativity either. Einstein showed that his field equations imply that a test-particle in an otherwise empty universe has inertial properties (Sciama [70] and Reinhardt [75]). The concept of inertial mass is as intrinsic to the body in general relativity as it was in Newtonian mechanics. Einstein did not succeed in constructing a theory where all the inertia of a body comes from its gravitational interaction with other bodies in the universe, in such a way that a body in an otherwise empty universe would have no inertia. Even his introduction of the cosmological term in general relativity did not provide a remedy, because in 1917 de Sitter found a solution of his modified field equations in the absence of matter [85, p. 287]. Einstein could never avoid the appearance of inertia relative to space in his theories, although this was required by Mach's principle.

Erwin Schrödinger (1887-1961) presented another argument showing that general relativity does not comply with Mach's principle [125] and [83]. In this article he says: "The general theory of relativity too in its original form[2] could *not* yet satisfy the Machian requirement, as was soon recognized. After the secular precession of the perihelion of Mercury was deduced, in amazing agreement with experiment, from it, every naive person had to ask: With respect to *what*, according to the *theory*, does the orbital ellipse perform this precession, which according to *experience* takes place with respect to the average system of the fixed stars? The answer that one receives is that the theory requires this precession to take place with respect to a coordinate system in which the gravi-

---

[2]A. Einstein, *Ann. d. Phys.* 49. S. 769. 1916.

tational potentials satisfy certain boundary conditions at infinity. However, the connection between these boundary conditions and the presence of the masses of the fixed stars was in no way clear, since these last were not included in the calculation at all." As the fixed stars were not included in the calculations of the precession of the perihelion of the planets in general relativity, it does not make sense to say that this precession is relative to the stars. On the other hand, the observations made by astronomers indicate that this precession happens relative to the background of fixed stars. This can only be a coincidence in general relativity. In relational mechanics this will no longer be a coincidence. It will be shown that it is the distant universe which generates the inertial force $-m\vec{a}$ or the kinetic energy $mv^2/2$. The distant universe has a fundamental influence over the bodies of the solar system. The precession of the perihelion calculated with relational mechanics is really relative to the distant universe, and not relative to an abstract frame disconnected from the distant matter in the cosmos.

Other discussions showing that Einstein's general theory of relativity does not implement Mach's principle can be found in Jammer [33, pp. 194-199], Ghins [62] and Borzeszkowski and Treder [126].

All of this shows that even in Einstein's general theory of relativity the concepts of absolute space or preferred inertial systems of reference disconnected from the distant matter are still present. The same happens with the inertia or with the inertial mass of bodies.

7.3.4 Newton's Bucket Experiment

How does Einstein's general theory of relativity cope with Newton's key bucket experiment? As before, let us concentrate on two situations. In the first one the water and the bucket are at rest relative to the earth, and in the second situation both are spinning together with a constant angular velocity ω_b relative to the earth. As $\omega_b \gg \omega_e \gg \omega_s \gg \omega_g$, during this experiment we can treat the earth as essentially without rotation relative to the frame of fixed stars and also relative to the frame of distant galaxies. Here $\omega_e \approx 7 \times 10^{-5}$ s^{-1} is the angular rotation of the earth relative to the fixed stars each day, $\omega_s \approx 2 \times 10^{-7}$ s^{-1} is the angular rotation of the solar system relative to the fixed stars with a period of one year and $\omega_g \approx 8 \times 10^{-16}$ s^{-1} is the angular rotation of the solar system around the center of our galaxy relative to the frame of distant galaxies with a period of 2.5×10^8 years.

As we have seen, the force exerted on the water molecules by the bucket is the same in both situations, as they are at rest relative to one another in both cases. This means that, in general relativity also, the bucket is not responsible for the concave form of the water surface.

In general relativity, the force exerted by the earth on the water in the first situation is essentially the Newtonian result of the weight of the water pointing vertically downwards. This will not be appreciably modified due to the rotation of the water relative to the earth in the second situation, as $v_w \ll c$, where v_w is the tangential velocity of any water molecule relative to the earth. In other words, as the velocities involved in this problem are negligible compared with light velocity, relativistic corrections will not be involved (they will not be of any importance). This means that in general relativity the rotation of the water relative to the earth cannot be responsible for the concavity of the water.

What about the fixed stars and distant galaxies? As we have seen, Mach believed the answer of the puzzle lay in the rotation of the water relative to distant matter. But in general relativity, there are no observable effects in a laboratory from a spherically symmetric agglomeration of matter at rest around it. In general relativity, the fixed stars and the distant galaxies exert essentially zero net force on any molecule of water in the first situation, as they are more or less evenly distributed around the earth. In the second situation seen from the earth, the same thing happens, as we now have the water moving relative to the fixed distant bodies. Consequently the fixed stars and distant galaxies do not exert any force, such as $-m\vec{a}$, on the water molecules. The consequence of this is that in general relativity the concave form of the water surface in the second situation is not due to its rotation relative to the bucket, nor relative to the fixed stars and distant galaxies.

The consequence of all this is that the concave form of the water must, according to general relativity, be due to its rotation relative to something else disconnected from matter. It might be Newton's absolute space or an inertial frame of reference which is completely disconnected (without any physical relation to) from the distant matter in the cosmos. Once more we see that general relativity retains the Newtonian concepts of absolute space and absolute motion (or, if you prefer, the concept of an inertial frame disconnected from distant matter).

To emphasize this point, suppose we are in an inertial frame of reference analysing the rotation of the water together with the bucket (second situation described above). In practice we know that in inertial frames the distant galaxies are essentially without translational acceleration and without rotation. This is a coincidence in Newtonian mechanics and also in general relativity, as there are no connections between these two entities (inertial frames and distant galaxies) according to these theories. To simplify the analysis, let us suppose we are in an inertial frame in which the distant galaxies homogeneously distributed in the sky are at rest without rotation. In this frame the bucket and the water are spinning together and the water surface is concave. As there are no observable effects in general relativity from a spherically symmetric distribution of matter, we can double the number and amount of matter of the galaxies around the bucket without influencing the concavity of the water surface. Alternatively, we could make all the distant galaxies disappear (literally annihilate them from the universe) without the slighest difference in the shape of the water surface. This is in complete disagreement with Mach's ideas (that the concavity of the water was due to its rotation relative to distant matter). This means that according to Mach's ideas, if the distant matter disappears, the concavity of water should vanish accordingly. Or, if we double the amount of distant matter, the concavity of water should double for the same relative rotation. None of this happens in Einstein's general theory of relativity.

But the situation becomes hopeless in the frame of reference O' which rotates with the bucket and with the water in the second situation. Now we have the bucket and the water at rest in this new frame, despite the concave form of the water surface. In Newtonian mechanics the term $m_i \vec{a}$ describing the motion of the water and responsible for the concavity of the surface in the previous frame of reference O becomes zero in this new frame O', as the water is now seen at rest. As $\vec{a}' = 0$ we get $m_i \vec{a}' = 0$. But according to Newtonian mechanics in the frame O', a centrifugal force $m_i \vec{\omega} \times (\vec{\omega} \times \vec{r}')$ acts on the water. This centrifugal force has exactly the same value $m_i \vec{a}$ had in the previous frame of reference. We may also say that the term $m_i \vec{a}$ has been transformed into the centrifugal force. The centrifugal force thus has exactly the right value to deform the water surface by the same amount as in the previous frame of reference O. Hence, a quantitative explanation is still possible in Newtonian mechanics not only in the inertial frame O (utilizing $m_i \vec{a}$), but also in the rotating frame O' (utilizing the centrifugal force). But in Eintein's general theory of relativity a strange thing happens. Although the fixed stars and distant galaxies exerted no force on the water in the frame O in which the stars and distant galaxies were seen at rest, the same does not happen in this frame O' of the bucket in which the stars and galaxies are seen rotating with $\vec{\omega}_{so'} = -\vec{\omega}_{bo}$, where $\vec{\omega}_{bo}$ is the angular rotation of the bucket and water relative to O. Now, due to the Thirring's force, (7.1), there will appear a real gravitational force exerted by the spinning distant matter on the water. This force did not exist in the frame of reference O. The problem is that this new force is not exactly the Newtonian fictitious centrifugal force. In it appears the new axial term (proportional to $(\vec{\omega} \cdot \vec{r})\vec{\omega}$) which has no analogue in Newtonian theory.

We saw previously that in general relativity, if we are in an inertial frame of reference O the concavity of water will be independent of the amount of distant matter around the bucket. But here we see that if we are in the frame O' rotating with the bucket and water, so that the set of distant galaxies is rotating in the opposite direction, the distant matter will exert a real gravitational force on the water given by Thirring's expression. In this frame of reference the galaxies influence the motion of the water and the shape of its surface. If we double the number of distant galaxies, the concavity of the water will change accordingly!

This is an undesirable consequence, as the physical situation is always the same, only seen from different frames of reference. It does not make sense for the galaxies to exert real gravitational forces on the water in one frame (with possible physical consequences, such as changing the form or concavity of its surface) and none at all in the other frame. In the frame O we can double or eliminate with the distant galaxies without changing the concavity of the water, while in the frame O' there is an influence (although we are analysing the same situation only from a different perspective): if we double the number of galaxies the water can may even overflow the bucket!

In Newtonian mechanics the situation was much better and more coherent. Whether distant matter was at rest or rotating, it never exerted any net force on the water. We could explain the concavity of the water in the inertial system O utilizing $m_i \vec{a}$, or in the frame O' rotating with the water introducing the centrifugal force $m_i \vec{\omega} \times (\vec{\omega} \times \vec{r}')$ ($m_i \vec{a}$ of frame O was transformed into $m_i \vec{\omega} \times (\vec{\omega} \times \vec{r}')$ in frame O'). Neither the centrifugal force nor $m_i \vec{a}$ had any relation to the distant galaxies. But in general relativity we have a gravitational frame-dependent force. In other words, the gravitational force between material bodies (between the water and distant galaxies here) depends on the state of motion of the observer. When the distant galaxies are seen at rest relative to O and the water rotates relative to them, they do not influence the concavity of the water surface, so that even when they disappear or are doubled in number, the concavity will be the same. But when we see the galaxies in the frame O' rotating in the opposite direction, while the bucket and water appear at rest, then according to Thirring's expression, there will be a real gravitational influence of the distant galaxies on the water. This means that in this frame O' the degree of concavity (whether or not the water overflows the bucket) is a function of the mass of distant galaxies! This result is certainly undesirable in any physical theory.

The same thing will happen according to general relativity in Newton's two globes experiment. In the frame of distant galaxies the tension in the cord is independent of the number of galaxies, while in the frame which rotates with the globes, the tension in the cord will be a function of the mass of distant galaxies due to Thirring's force.

It might be thought that this is a negligible effect, but this is not the case. When we integrate Thirring's force over the whole known universe we obtain an expression of the same order of magnitude as the Coriolis and centrifugal forces of classical mechanics. Replacing M by $dM = 4\pi R^2 \rho dR$ in Eq. (7.1) and integrating from zero to Huble's radius $R_o = c/H_o$ yields:

$$\vec{F} = -\frac{8G\rho}{15H_o^2} \left[m\vec{\omega} \times (\vec{\omega} \times \vec{r}) + 10m\vec{v} \times \vec{\omega} + 2m(\vec{\omega} \cdot \vec{r})\vec{\omega} \right] . \tag{7.2}$$

To arrive at this result we supposed a constant matter density ρ. The coefficient $8G\rho/15H_o^2$ is of the order of unity with the known values of G, H_o and the mean matter density of the universe, $\rho \approx 10^{-27}$ kg/m^3.

Eq. (7.2) gives the gravitational force exerted by the spinning universe on any body, according to general relativity. It has the same order of magnitude as the classical Coriolis and centrifugal forces. But the form and numerical values of Thirring's force are different from the classical ones. This means that Foucault's pendulum or the flattening of the earth, when analysed from the earth's frame of reference in which the distant galaxies are seen as rotating, should, according to general relativity, have values different from those observed experimentally. This is one of the main quantitative flaws of general relativity.

Let us show this in detail. In the earth's non-inertial frame of reference S' general relativity yields the same equation of motion as classical mechanics for test bodies moving with small velocities compared with light velocity, Eq. (3.17):

$$m\frac{d^2\vec{r}\,'}{dt^2} = \vec{F} - m\vec{\omega} \times (\vec{\omega} \times \vec{r}\,')$$

$$- 2m\vec{\omega} \times \frac{d\vec{r}\,'}{dt} - m\frac{d\vec{\omega}}{dt} \times \vec{r}\,' - m\frac{d^2\vec{h}}{dt^2} \, . \tag{7.3}$$

The centrifugal and Coriolis forces are not due to interactions with the distant universe. However, in general relativity we must include in $\vec{F}$ not only the local forces like the earth's gravity acting on the test body but also the real gravitational force exerted by the spinning distant universe acting on the test body. This force exerted by the spinning distant universe is given by Eq. (7.2). To have an order of magnitude of this force we utilize the critical density characterized in general relativity, namely: $\rho = 3H_o^2/8\pi G$. This means that the coefficient in front of the square brackets will have the typical value of $1/15$. Performing the calculations as in chapter 3 yields a flattening of the earth given by $R_>/R_< \approx 1 + (1 + 1/15)5\omega^2 R^3/4GM$, while in Newtonian mechanics it is given by $1 + 5\omega^2 R^3/4GM$. The plane of oscillation of Foucault's pendulum, on the other hand, will precess at a rate given by $\Omega = (1 + 1/3)\omega \sin \alpha$, where α is the local latitude. The factor $1/3$ was due to $5 \times 1/15$, as in Eq. (7.2) the analogous to Coriolis force is 5 times larger than the analogous to the centrifugal one. However, what is observed experimentally is a precession given by Newtonian mechanics, namely: $\Omega = \omega \sin \alpha$. These two numerical calculations show that general relativity does not yield the measured values of these quantities in the earth's frame of reference.

This analysis shows clearly that in general relativity, kinematically equivalent situations are not dynamically equivalent. Mach, on the other hand, believed it would be possible to formulate a mechanics in which this would be accomplished. Once more we see that Einstein's theory does not implement Mach's ideas.

The discussion of this section shows that general relativity cannot cope with Newton's bucket or two globes experiments in all frames of reference. Classical Newtonian mechanics, on the other hand, could explain these two experiments in all frames of reference, with $m_i \vec{a}$ in the inertial ones or with the fictitious centrifugal forces in the non-inertial ones. Neither of these ($m_i \vec{a}$ or the centrifugal force) is related to distant matter, which shows the coherence of the theory. Einstein's theory does not present the same coherence.

There are other problems with Einstein's general theory of relativity: *e.g.* the inertial mass is not well defined and it does not comply with the principle of the conservation of energy [127]. We will not go into further detail here.

7.4 General Comments

In conclusion we may say that there are many problems with Einstein's special and general theories of relativity. We stress some of them here.

1) They are based on Lorentz's formulation of electrodynamics, which suffers from asymmetries pointed out by Einstein and many others. These asymmetries do not appear in the observed phenomena of induction. There is another theory of electrodynamics which naturally avoids all these asymmetries, namely, Weber's electrodynamics. A complete discussion of this theory with many recent references can be found elsewhere ([93] and [12]). In order to explain inertia, Weber's law is a better starting point than Lorentz's force.

2) Einstein's special theory of relativity maintains the concept of absolute space and of inertial frames disconnected from distant matter. Moreover, it introduces another absolute entity, namely, the velocity of light in vacuum. Nothing in physics leads to the conclusion that light velocity should be constant irrespective of the motion of the observer or of the detector. All velocities known to us are constant relative to the source (like bullets) or constant relative to the medium (like sound velocity which is constant relative to air, irrespective of the motion of the source). But all of them vary according to the motion of the observer or detector. To assert the opposite, as Einstein did, can only lead to the necessity of introducing strange and unnecessary concepts in physics such as time dilation, contraction of lengths, proper times *etc.* Einstein's theory maintained the concept of inertial frames disconnected from distant matter and introduced the absolute character of light velocity in vacuum. To avoid confusion with Einstein's theories of relativity we adopt the name "Relational Mechanics" for the theory developed here. Our work is based only on relative concepts, without absolute space, absolute time, inertial mass, inertial frames of reference or absolute light velocity.

3) Einstein begins to interpret the velocity in Lorentz's force as the velocity of the test charge relative to the observer (and not to the dielectric in which the charge is moving, nor to a special frame like the ether, nor to the magnet or current-carrying wire generating the magnetic field). This Einsteinian point of view is contrary to the interpretations of Thomson, Heaviside and Lorentz. No experiment forced this new interpretation of the terms appearing in the basic force law. The induction of currents discussed by Einstein was known since 1831, while Thomson's paper is from 1881. Thomson (1881), Heaviside (1889) and Lorentz (1895) maintained different interpretations although dealing with the same experiments.

4) Einstein correctly pointed out that the best way to implement Mach's principle was to utilize only the distance between interacting bodies and their relative velocities and accelerations. Unfortunately he himself did not follow this route because he thought it was impractical. He was mistaken in this regard, as we show in this book.

5) He correctly pointed out four features which should be implemented in any model designed to incorporate Mach's principle. His own general theory of relativity does not completely reproduce these four elements, as he himself concluded and as we have showed in this book. As we will see, these four consequences follow directly and quantitatively from a relational mechanics based on the works of Mach and Weber.

6) The forces similar to the centrifugal and Coriolis forces which appear in general relativity with Thirring's force are not as expected. The numerical coefficients are not exactly equivalent to the terms which we know to exist in non-inertial frames of reference of classical mechanics. The force analogous to the Coriolis force, in particular, appears five times larger than expected, assuming the centrifugal force to be correct; see Eq. (7.1). Moreover, there appear spurious terms such as the axial terms, which we cannot get rid of. It is known that these axial terms do not exist. In other words, no one has ever found any effect or force in non-inertial frames which pointed in the direction of $\vec{\omega}$.

7) General relativity cannot explain Newton's bucket experiment in all frames of reference, contrary to what happens in classical mechanics.

8) The only frame-dependent forces in Newtonian mechanics were the inertial forces ($m_i \vec{a}$, centrifugal, Coriolis, *etc.*). In Newtonian mechanics these forces have no relation to the fixed stars or distant matter in the universe. For this reason it was understandable that they had this odd property. All other forces between material bodies were relational forces, depending only on intrinsic quantities of the system, such as the distance or velocity between material bodies. Examples include: Newton's law of gravitation, the elastic force of a spring, the force of friction in a fluid which depends on the relative velocity between the test body and the surrounding medium, Coulomb's force, contact

forces *etc.* Einstein changed all this by introducing frame-dependent electromagnetic forces with his new interpretation of the velocity in Lorentz's force law. He also introduced a frame-dependent gravitational force with his general theory of relativity, as we saw when discussing Newton's bucket experiment in this theory.

In our view, the theoretical concepts of length contraction, time dilation, Lorentz invariance, Lorentz's transformations, covariant and invariant laws, Minkowski metric, four-dimensional space-time, energy-momentum tensor, Riemannian geometry applied to physics, Schwarzschild line element, tensorial algebras in four-dimensional spaces, quadrivectors, metric tensor $g_{\mu\nu}$, proper time, contravariant four-vectors and tensors, geodetic lines, Christoffel symbols, super strings, curvature of space, *etc.* have the same role as the epicycles in the Ptolemaic theory.

Although Einstein was greatly influenced by Mach's ideas, Mach himself rejected Einstein's theories of relativity. This can be seen from his statement in the preface of his last book *The Principles of Physical Optics - An Historical and Philosophical Treatment* [128]. There he says that he was compelled to cancel his contemplation of the relativity theory (which he finds to be growing more and more dogmatical) and that his disclaims to be a forerunner of the relativists.

Additional proofs that Mach opposed Einstein's theories of relativity can be found in Mach's biography by Blackmore [63], and in his important paper "Ernst Mach leaves 'The Church of Physics' " [129].

Part II

New World

Chapter 8

Relational Mechanics

8.1 Basic Concepts and Postulates

We now present a new mechanics to replace the Newtonian and Einsteinian mechanics. We call it "Relational Mechanics." We begin with the complete formulation of the theory, and then discuss its applications. In the final chapter we outline the history of relational mechanics, highlighting the main developments and putting everything in perspective.

By relational mechanics we understand a formulation of mechanics (the study of the equilibrium and motions of masses) only in terms of relative quantities, avoiding the use of absolute concepts, such as Newton's absolute space and time. In relational mechanics we also do not utilize quantities which depend on the observer, such as the velocity in Lorentz's force, as interpreted by Einstein. We do not utilize the older expression "relativistic mechanics" to avoid confusion with Einstein's special and general theories of relativity.

We begin presenting some basic (or primitive) concepts necessary to define more complex ones. We do not define these basic concepts, since we wish to avoid vicious circles. The basic or primitive concepts which we will need are: (1) gravitational mass, (2) electrical charge, (3) distance between material bodies, (4) time between physical events, and (5) force or interaction between material bodies.

It may be possible to derive the gravitational force from an electromagnetic force, as we have shown elsewhere ([130] and [131]). If this is the case, gravitational mass will not be a basic concept. As this is not yet proved, we will continue treating it as one.

At no time do we introduce the concepts of inertia or inertial mass, inertial frames of reference, or the concepts of absolute space and absolute time.

We now present the three postulates of relational mechanics:

(A) Force is a vectorial quantity describing the interaction between material bodies.

(B) The force that a point particle A exerts on a point particle B is equal and opposite to the force that B exerts on A, and is directed along the straight line connecting A to B.

(C) The sum of all forces of any nature (gravitational, electric, magnetic, elastic, nuclear, *etc.*) acting on any body is always zero in all frames of reference.

The first postulate qualifies the nature of a force (stating that it is a vectorial quantity, with magnitude and direction). With this postulate we are assuming the law of the parallelogram of forces (that they add like vectors). Observe only that we are not yet talking about accelerations, only forces. It must also be clear that force is an interaction between material bodies. For instance, it does not describe an interaction of a body with "space."

The second postulate is similar to Newton's action and reaction law, namely: $\vec{F}_{AB} = -\,\vec{F}_{BA}$. In addition, we are specifying that all forces between point particles, no matter what their origin (electrical, elastic, gravitational, chemical, nuclear *etc.*) are directed along the straight line connecting these bodies. It is important to emphasize here the notion of "point" particles. The reason is simple and can be illustrated as follows: Consider an electric dipole made up of two point charges $q_1 > 0$ and $-q_1$ separated by a distance d_1. We choose a frame of reference O with origin at the center of this dipole, with z axis along the line connecting q_1 to $-q_1$, pointing from $-q_1$ to q_1. The electric dipole moment $\vec{p}_1$ is defined by $\vec{p}_1 \equiv q_1 d_1 \hat{z}$. Consider another point charge $q_2 > 0$ located along the x axis, at a distance r_2 from the origin. We will consider all charges at rest, so that this is a simple electrostatic problem. The force exerted by q_1 on q_2 is along the straight line connecting them. The force exerted by $-q_1$ on q_2 is along the straight line connecting these two charges. Adding these two expressions yields the resultant force exerted by the dipole $\vec{p}_1$ on q_2. This force is along the z axis, as in Figure 8.1.

Even when $d_1 \ll r_2$ the force between the dipole and q_2 is not along the x axis, which might considered the straight line connecting the "point" dipole (its center) to a far away q_2. The reason for this behaviour is that even in this case in which $d_1 \ll r_2$ the dipole is not really a point, as there is a small distance between its two charges.

Neglecting cases like this, it is often possible to replace two large bodies A and B by point particles when their dimensions (averate or maximal diameters) are much smaller than the distance between their centers.

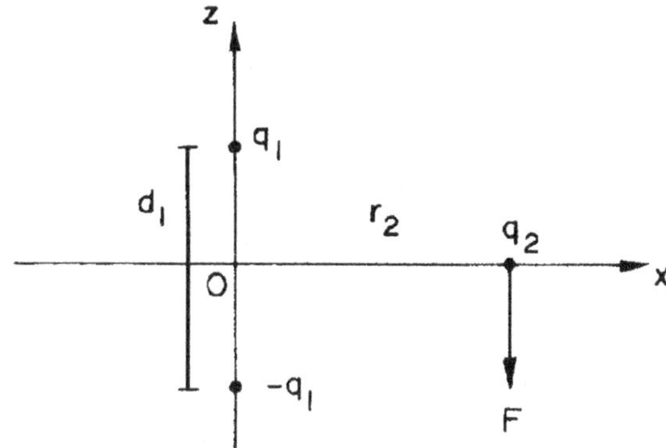

Figure 8.1: Point charge interacting with an electric dipole.

The third postulate is the main departure from Newtonian mechanics. We may call it the principle of dynamical equilibrium. It states that the sum of all forces on a body is always zero, even when the test body is in motion and accelerated relative to another body, to ourselves or to any other frame of reference. Later on we *derive* a law similar to Newton's second law of motion.

The advantage of this postulate compared to Newton's second law of motion is that we do not introduce the concepts of inertia, inertial system of reference, inertial mass and absolute space. In Newtonian mechanics, the sum of all forces was equal to the time variation of linear momentum (inertial mass times velocity). For constant mass, this was equal to the inertial mass of the test body times its acceleration relative to absolute space or to an inertial frame of reference. This means that these concepts had to be introduced and clarified beforehand, and were an essential part of Newton's second law. This is the greatest advantage of our third postulate. Moreover, it is valid in all frames of reference, while the Newtonian law was valid only in inertial frames of reference (in non-inertial frames we needed to utilize fictitious forces in Newtonian mechanics). Suppose a person on the earth's surface throws a rock upwards in the presence of a strong wind affecting the rock's motion (influencing its direction and velocity). The person will apply the postulate that the resultant force acting on the rock is zero, even when the rock is rising, falling, stopping at the floor and staying there at rest. In the frame of the rock (a frame that is always at rest relative to the rock) we should also apply the postulate that the resultant force acting on it is always zero. In any other arbitrary frame of reference which is in motion relative to the earth or the rock, the postulate that the resultant force acting on the rock is zero at all times should also be applied.

When we say that the sum of all forces on any body is always zero in all frames of reference, we arrive at another result, in agreement with Mach's ideas.

We can multiply all forces by the same constant (no matter its dimensional units) without affecting the results. The only thing that will matter is the ratio of any two forces. We can never know the absolute value of any force, only how much one force is larger or smaller than another. The units of the forces also remain unspecified, provided all forces have the same unit.

If we are working with energies instead of forces, these three postulates might be replaced by one simple postulate:

> The sum of all interaction energies (gravitational, electromagnetic, elastic, nuclear, *etc.*) between any body and all other bodies in the universe is always zero in all frames of reference.

Once more, only the ratio of energies will be important. This postulate may be called the principle of the conservation of energy. The advantage of this postulate over the analogous postulate in classical mechanics (the sum of the kinetic and potential energies is a constant) is that we do not introduce the concept of kinetic energy, $m_i v^2/2$. This kinetic energy has embedded in it the concepts of inertial mass (m_i) and absolute space or inertial systems (the frames where the velocity v is to be measured). Later on, we *derive* an analogue to this classical kinetic energy. We also *derive* a theorem for the conservation of energy analogous to the one of classical mechanics.

8.2 Electromagnetic and Gravitational Forces

These postulates refer only to the forces between interacting bodies. Until now the concepts of gravitational mass, charge and distance between bodies have not appeared. In order to implement these postulates and obtain the equations of motion following Mach's ideas, we need some expressions for the forces and energies. The postulates only make sense together with specific forces and energies describing the several types of interactions, as in the case of Newton's laws of motion. Here we introduce the main contribution of Wilhelm Weber (1804-1891). In 1848 he proposed that the energy of interaction between two electrical charges q_1 and q_2 be given by [12, Chapter 3]:

$$U_{12} = H_e q_1 q_2 \frac{1}{r_{12}} \left(1 - \frac{\dot{r}_{12}^2}{2c^2} \right) . \tag{8.1}$$

Here r_{12} is the distance between the charges and $\dot{r}_{12} \equiv dr_{12}/dt$ is their radial relative velocity and $c = 3 \times 10^8$ m/s, while H_e is a constant which depends on the system of units. In Newtonian mechanics and in the International System of Units it is written as $1/4\pi\varepsilon_o$. In relational mechanics its value (or its ratio to H_g, see below) will be specified later.

The force exerted by 2 on 1 can be obtained by $\vec{F}_{21} = -\hat{r}_{12} dU_{12}/dr_{12}$, yielding the result Weber had already proposed in 1846, namely:

$$\vec{F}_{21} = H_e q_1 q_2 \frac{\hat{r}_{12}}{r_{12}^2} \left(1 - \frac{\dot{r}_{12}^2}{2c^2} + \frac{r_{12}\ddot{r}_{12}}{c^2} \right) . \tag{8.2}$$

Weber had arrived at this result in order to unify electrostatics (Coulomb's force) with electrodynamics (Ampère's force between current elements) and Faraday's law of induction [12, Chapter 3].

The charges q_1 and q_2 are located at the position vectors $\vec{r}_1 \equiv x_1\hat{x}+y_1\hat{y}+z_1\hat{z}$ and $\vec{r}_2 \equiv x_2\hat{x}+y_2\hat{y}+z_2\hat{z}$ relative to the origin of an arbitrary frame of reference (not necessarily inertial in the Newtonian sense). The unit vectors $\hat{x}$, $\hat{y}$ and $\hat{z}$ point along the orthogonal axes x, y and z of this frame of reference. Their velocities and accelerations relative to the origin of this coordinate system are given by: $\vec{v}_1 = d\vec{r}_1/dt = \dot{x}_1\hat{x} + \dot{y}_1\hat{y} + \dot{z}_1\hat{z}$, $\vec{v}_2 = d\vec{r}_2/dt = \dot{x}_2\hat{x} + \dot{y}_2\hat{y} + \dot{z}_2\hat{z}$, $\vec{a}_1 = d^2\vec{r}_1/dt^2 = d\vec{v}_1/dt = \ddot{x}_1\hat{x}+\ddot{y}_1\hat{y}+\ddot{z}_1\hat{z}$, $\vec{a}_2 = d^2\vec{r}_2/dt^2 = d\vec{v}_2/dt = \ddot{x}_2\hat{x}+\ddot{y}_2\hat{y}+\ddot{z}_2\hat{z}$. The position vector of one charge relative to another, and their relative vectorial velocity and acceleration in this frame of reference, are given by, respectively:

$$\vec{r}_{12} \equiv \vec{r}_1 - \vec{r}_2 = (x_1 - x_2)\hat{x} + (y_1 - y_2)\hat{y} + (z_1 - z_2)\hat{z} \equiv x_{12}\hat{x} + y_{12}\hat{y} + z_{12}\hat{z} ,$$

$$\vec{v}_{12} \equiv \frac{d\vec{r}_{12}}{dt} = \vec{v}_1 - \vec{v}_2 = (\dot{x}_1 - \dot{y}_1)\hat{x} + (\dot{y}_1 - \dot{y}_2)\hat{y} + (\dot{z}_1 - \dot{z}_2)\hat{z}$$

$$\equiv \dot{x}_{12}\hat{x} + \dot{y}_{12}\hat{y} + \dot{z}_{12}\hat{z} ,$$

$$\vec{a}_{12} \equiv \frac{d^2\vec{r}_{12}}{dt^2} = \frac{d\vec{v}_{12}}{dt} = \vec{a}_1 - \vec{a}_2 = (\ddot{x}_1 - \ddot{x}_2)\hat{x} + (\ddot{y}_1 - \ddot{y}_2)\hat{y} + (\ddot{z}_1 - \ddot{z}_2)\hat{z}$$

$$\equiv \ddot{x}_{12}\hat{x} + \ddot{y}_{12}\hat{y} + \ddot{z}_{12}\hat{z} .$$

They are separated by a distance

$$r_{12} \equiv |\vec{r}_1 - \vec{r}_2| = [(x_1 - x_2)^2 + (y_1 - y_2)^2 + (z_1 - z_2)^2]^{1/2}$$

$$= \sqrt{x_{12}^2 + y_{12}^2 + z_{12}^2} .$$

The unit vector pointing from q_2 to q_1 is given by:

$$\hat{r}_{12} \equiv \frac{\vec{r}_{12}}{r_{12}} .$$

Their relative *radial* velocity and acceleration are given by, respectively:

$$\dot{r}_{12} \equiv \frac{dr_{12}}{dt} = \frac{x_{12}\dot{x}_{12} + y_{12}\dot{y}_{12} + z_{12}\dot{z}_{12}}{r_{12}} = \hat{r}_{12} \cdot \vec{v}_{12} \; ,$$

$$\ddot{r}_{12} \equiv \frac{d\dot{r}_{12}}{dt} = \frac{d^2 r_{12}}{dt^2} = \frac{\vec{v}_{12} \cdot \vec{v}_{12} - (\hat{r}_{12} \cdot \vec{v}_{12})^2 + \vec{r}_{12} \cdot \vec{a}_{12}}{r_{12}} \; .$$

The many properties and advantages of Weber's electromagnetic theory have been discussed at length in a separate book, *Weber's Electrodynamics* [12].

By analogy with Weber's electrodynamics, we propose as the basis of relational mechanics that Newton's law of gravitation be modified in accordance with Weber's law. In particular, the energy of interaction between two gravitational masses m_{g1} and m_{g2} and the force exerted by 2 on 1 should be given by:

$$U_{12} = -H_g \frac{m_{g1}m_{g2}}{r_{12}} \left(1 - \xi \frac{\dot{r}_{12}^2}{2c^2} \right) \; , \tag{8.3}$$

$$\vec{F}_{21} = -H_g m_{g1} m_{g2} \frac{\hat{r}_{12}}{r_{12}^2} \left[1 - \frac{\xi}{c^2} \left(\frac{\dot{r}_{12}^2}{2} - r_{12}\ddot{r}_{12} \right) \right] \; . \tag{8.4}$$

In these equations we assume H_g and ξ to be constants. With $\xi = 0$ or $c \rightarrow \infty$ we recover the usual potential energy and force of Newtonian mechanics, if we put $H_g = G$. For the time being we only require that $\xi > 0$. Later on we will find that $\xi = 6$ in order to derive the observed precession of the perihelion of the planets.

In order to avoid the gravitational paradox presented previously, and an analogous one which appears when we implement Mach's principle with relational mechanics, we can utilize the following modifications of these energies and forces:

$$U_{12} = -H_g \frac{m_{g1}m_{g2}}{r_{12}} \left(1 - \xi \frac{\dot{r}_{12}^2}{2c^2} \right) e^{-\alpha r_{12}} \; , \tag{8.5}$$

$$\vec{F}_{21} = -H_g m_{g1} m_{g2} \frac{\hat{r}_{12}}{r_{12}^2} \left[1 - \frac{\xi}{c^2} \left(\frac{\dot{r}_{12}^2}{2} - r_{12}\ddot{r}_{12} \right) \right.$$

$$\left. + \alpha r_{12} \left(1 - \frac{\xi}{2} \frac{\dot{r}_{12}^2}{c^2} \right) \right] e^{-\alpha r_{12}} \; . \tag{8.6}$$

In these equations α is a constant with dimensions of *length*$^{-1}$.

The force is also derived from

$$\vec{F}_{21} = -\hat{r}_{12}\frac{dU_{12}}{dr_{12}} \ .$$

The main properties of Weber's potential energy and force, as applied to electromagnetism and gravitation, are the following:

A) These forces follow the second postulate strictly, as they obey the law of action and reaction and are along the line connecting the interacting bodies.

B) We recover Coulomb's force and Newton's law of gravitation when there is no motion between the particles, when $\dot{r}_{12} = 0$ and $\ddot{r}_{12} = 0$. This will happen when the distance between the particles is a constant, even if they are moving together relative to an arbitrary frame of reference or to other bodies.

C) The most important property is that these energies and forces depend only on the relative distance, radial velocity and radial acceleration between the interacting particles. Although the position, velocity and acceleration of one particle relative to a frame of reference O may be different from the position, velocity and acceleration of the same particle relative to another frame of reference O', the relative distance, relative radial velocity and relative radial acceleration between the two particles are the same in both frames [12, Section 3.2]. In other words, these forces and energies are completely relational in their nature. They have the same values for all observers, irrespective of whether the observer is inertial from the Newtonian point of view.

All energies and force laws to be proposed in the future must have this property in order to implement Mach's principle. As we have shown before, Mach emphasized that "all masses and all velocities, and consequently all forces, are relative."

Even when we have a medium, as in the frictional force acting between a projectile and the surrounding air or water, only relational quantities should appear. For instance, the force of dynamic friction must be written in terms of the relative velocity between the projectile and the medium (air or water in this case). If one day the ether is found, the same must be true for it. The force between the ether and the particles must depend only on the relative velocity and acceleration between each particle and the ether, but not on the velocity and acceleration of the particles relative to any observer or frame of reference.

The situation in physics nowadays is quite different. In Newton's second law of motion we have accelerations relative to absolute space or to inertial frames of reference, and not relative to the bodies with which the test body is interacting. In Lorentz's force ($\vec{F} = q\vec{E} + q\vec{v} \times \vec{B}$) the velocity $\vec{v}$ is understood, after Einstein, as the velocity of the test charge q relative to the observer and not relative to the magnet or current-carrying wire with which it is obviously interacting.

8.3   Spherical Shell Interacting with a Particle

We first consider a test particle of gravitational mass m_g inside a spherical shell of gravitational mass dM_g. The shell has a radius R, thickness dR and an isotropic gravitational matter density ρ_g. We consider the shell to be stationary in a frame of reference U, with its center at the origin of U. The point mass m_g is located at $\vec{r}_{mU}$ and moves with velocity $\vec{v}_{mU} = d\vec{r}_{mU}/dt$ and acceleration $\vec{a}_{mU} = d^2\vec{r}_{mU}/dt^2$ relative to the origin U.

The gravitational mass of the spherical shell is given by:

$$dM_g = 4\pi\rho_g R^2 dR \ .$$

We now integrate Weber's gravitational energy of interaction between this test particle and the shell, Eq. (8.3), obtaining:

$$dU_{Mm}(r_{mU} < R) = -4\pi H_g m_g \rho_g R dR \left(1 - \frac{\xi}{6}\frac{\vec{v}_{mU}\cdot\vec{v}_{mU}}{c^2} \right) \ . \qquad (8.7)$$

In classical mechanics, the terms with $\vec{v}_{mU}$ would not appear. Only the constant term $-4\pi H_g m_g \rho_g R dR$ would be present after an analogous integration. But as we will see, it is the velocity term which will generate an analogue to the kinetic energy and an implementation of Mach's principle.

Integrating Eq. (8.4) to obtain the gravitational force exerted by the spinning shell on the internal particle yields:

$$d\vec{F}_{Mm}(r_{mU} < R) = -\frac{4\pi}{3} H_g \frac{\xi}{c^2} m_g \rho_g R dR \vec{a}_{mU} \ . \qquad (8.8)$$

This term would appear with Newton's law of gravitation. As we have seen, Newton proved that the gravitational force acting on a body inside a spherical shell is zero with his force. On the other hand, this term we have obtained here will be essential for the implementation of Mach's principle.

With the test particle localized outside the spherical shell, the interaction energy and the force exerted by the shell are given by:

$$dU_{Mm}(r_{mU} > R) = -H_g m_g (4\pi\rho_g R^2 dR)\frac{1}{r_{mU}}\left\{ 1 \right.$$

$$\left. - \frac{\xi(\hat{r}_{mU}\cdot\vec{v}_{mU})^2}{2c^2} - \frac{\xi}{6c^2}\frac{R^2}{r_{mU}^2}\left[\vec{v}_{mU}\cdot\vec{v}_{mU} - 3(\hat{r}_{mU}\cdot\vec{v}_{mU})^2\right] \right\} \ , \qquad (8.9)$$

$$d\vec{F}_{Mm}(r_{mU} > R) = -\frac{H_g m_g (4\pi\rho_g R^2 dR)}{r_{mU}^2}\left\{ \left[1 \right. \right.$$

$$+ \frac{\xi}{c^2}\left((\vec{v}_{mU} \cdot \vec{v}_{mU}) - \frac{3}{2}(\hat{r}_{mU} \cdot \vec{v}_{mU})^2 + \vec{r}_{mU} \cdot \vec{a}_{mU} \right) \right] \hat{r}_{mU}$$

$$+ \frac{\xi}{c^2}\frac{R^2}{r_{mU}^2}\left[\frac{r_{mU}}{3}\vec{a}_{mU} - (\hat{r}_{mU} \cdot \vec{v}_{mU})\vec{v}_{mU} \right.$$

$$\left. - \frac{\vec{v}_{mU} \cdot \vec{v}_{mU}}{2}\hat{r}_{mU} + \frac{5}{2}(\hat{r}_{mU} \cdot \vec{v}_{mU})^2\hat{r}_{mU} - (\vec{r}_{mU} \cdot \vec{a}_{mU})\hat{r}_{mU} \right] \right\} . \tag{8.10}$$

We now consider a test particle of gravitational mass m_g inside a spherical shell of gravitational mass dM_g. The shell has a radius R, thickness dR, an isotropic gravitational matter density ρ_g, and spins with an angular velocity $\vec{\omega}_{MS}(t)$ relative to an arbitrary frame of reference S. The center of the stationary (but spinning) shell is at the origin O of S.

The point mass m_g is located at $\vec{r}_{mS}$ and moves with velocity $\vec{v}_{mS} = d\vec{r}_{mS}/dt$ and acceleration $\vec{a}_{mS} = d^2\vec{r}_{mS}/dt^2$ relative to the origin O of S, as in Figure 8.2.

We integrate Weber's gravitational energy of interaction between this test particle and the shell, Eq. (8.3), obtaining [12, Chapter 7]:

$$dU_{Mm}(r_{mS} < R) = -4\pi H_g m_g \rho_g R dR$$

$$\times \left[1 - \frac{\xi}{6}\frac{(\vec{v}_{mS} - \vec{\omega}_{MS} \times \vec{r}_{mS}) \cdot (\vec{v}_{mS} - \vec{\omega}_{MS} \times \vec{r}_{mS})}{c^2} \right] . \tag{8.11}$$

In classical mechanics, the terms with $\vec{v}_{mS}$ and $\vec{\omega}_{MS}$ would not appear. Only the constant term $-4\pi H_g m_g \rho_g R dR$ would be present after an analogous integration. But as we will see, it is the velocity term which will generate an analogue to the kinetic energy and an implementation of Mach's principle.

Integrating Eq. (8.4) to obtain the gravitational force exerted by the spinning shell on the internal particle yields [12, Chapter 7]:

$$d\vec{F}_{Mm}(r_{mS} < R) = -\frac{4\pi}{3}H_g\frac{\xi}{c^2}m_g\rho_g R dR \left[\vec{a}_{mS} + \vec{\omega}_{MS} \times (\vec{\omega}_{MS} \times \vec{r}_{mS}) \right.$$

$$\left. + 2\vec{v}_{mS} \times \vec{\omega}_{MS} + \vec{r}_{mS} \times \frac{d\vec{\omega}_{MS}}{dt} \right] . \tag{8.12}$$

None of these terms would appear with Newton's law of gravitation. As we have seen, Newton proved that the gravitational force acting on a body inside

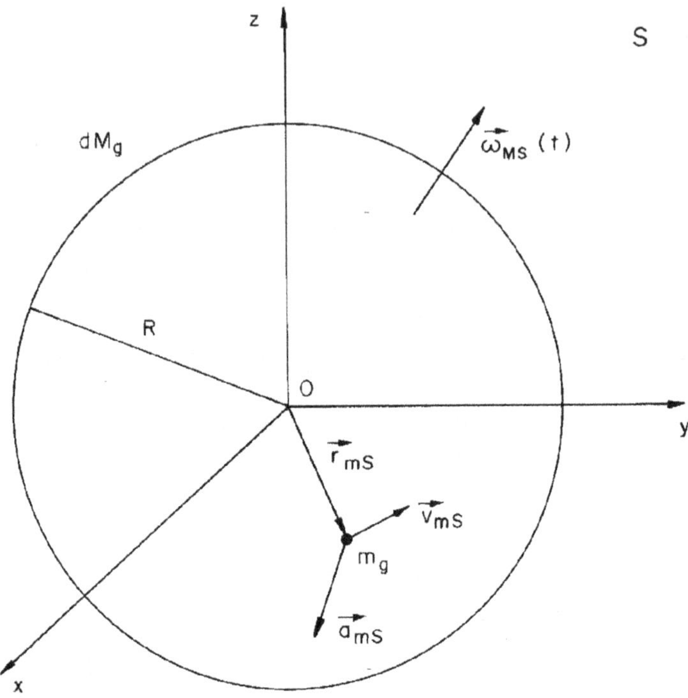

Figure 8.2: Spinning spherical shell of mass dM_g interacting with a material particle of mass m_g.

a spherical shell is zero with his force. On the other hand all the terms we have obtained here will be essential for the implementation of Mach's principle.

If the test particle were localized outside the spherical shell, the interaction energy and the force exerted by the shell would be given by [12]:

$$dU_{Mm}(r_{mS} > R) = -H_g m_g (4\pi\rho_g R^2 dR)\frac{1}{r_{mS}}\left\{1\right.$$

$$-\frac{\xi[\hat{r}_{mS}\cdot(\vec{v}_{mS} - \vec{\omega}_{MS}\times\vec{r}_{mS})]^2}{2c^2}$$

$$-\frac{\xi}{6c^2}\frac{R^2}{r_{mS}^2}\left[(\vec{v}_{mS} - \vec{\omega}_{MS}\times\vec{r}_{mS})\cdot(\vec{v}_{mS} - \vec{\omega}_{MS}\times\vec{r}_{mS})\right.$$

$$\left.\left. - 3[\hat{r}_{mS}\cdot(\vec{v}_{mS} - \vec{\omega}_{MS}\times\vec{r}_{mS})]^2\right]\right\}, \tag{8.13}$$

$$d\vec{F}_{Mm}(r_{mS} > R) = -\frac{H_g m_g (4\pi \rho_g R^2 dR)}{r_{mS}^2} \left\{ \left[1 \right. \right.$$

$$+ \frac{\xi}{c^2}\left((\vec{v}_{mS} \cdot \vec{v}_{mS}) - \frac{3}{2}(\hat{r}_{mS} \cdot \vec{v}_{mS})^2 + \vec{r}_{mS} \cdot \vec{a}_{mS} \right) \Big] \hat{r}_{mS}$$

$$+ \frac{\xi}{c^2}\frac{R^2}{r_{mS}^2}\left[\frac{r_{mS}}{3}\vec{a}_{mS} - (\hat{r}_{mS} \cdot \vec{v}_{mS})\vec{v}_{mS} \right.$$

$$- \frac{\vec{v}_{mS} \cdot \vec{v}_{mS}}{2}\hat{r}_{mS} + \frac{5}{2}(\hat{r}_{mS} \cdot \vec{v}_{mS})^2\hat{r}_{mS}$$

$$- (\vec{r}_{mS} \cdot \vec{a}_{mS})\hat{r}_{mS} + (\hat{r}_{mS} \cdot \vec{v}_{mS})(\vec{\omega}_{MS} \times \vec{r}_{mS})$$

$$+ \frac{2}{3}r_{mS}(\vec{v}_{mS} \times \vec{\omega}_{MS}) + \frac{r_{mS}}{3}(\vec{\omega}_{MS} \cdot \vec{r}_{mS})\vec{\omega}_{MS} + \frac{r_{mS}^2\omega_{MS}^2}{6}\hat{r}_{mS}$$

$$- \frac{(\vec{r}_{mS} \cdot \vec{\omega}_{MS})^2}{2}\hat{r}_{mS} + [\vec{r}_{mS} \cdot (\vec{\omega}_{MS} \times \vec{v}_{mS})]\hat{r}_{mS}$$

$$+ \frac{r_{mS}}{3}\left(\vec{r}_{mS} \times \frac{d\vec{\omega}_{MS}}{dt} \right) \Big] \Big\} \ . \tag{8.14}$$

Suppose now the spherical shell still spinning with $\vec{\omega}_{MS}$ relative to S. But now let us suppose the center of the shell is localized at $\vec{R}_{oS}$ and is moving with velocity $\vec{V}_{oS}$ and acceleration $\vec{A}_{oS}$ relative to the origin of S. Integration of Eqs. (8.3) and (8.4) gives results analogous to Eqs. (8.11) to (8.14), but with $\vec{r}_{mS} - \vec{R}_{oS}$, $|\vec{r}_{mS} - \vec{R}_{oS}|$, $\hat{r}_{mo} \equiv (\vec{r}_{mS} - \vec{R}_{oS})/|\vec{r}_{mS} - \vec{R}_{oS}|$, $\vec{v}_{mS} - \vec{V}_{oS}$ and $\vec{a}_{mS} - \vec{A}_{oS}$ instead of $\vec{r}_{mS}$, r_{mS}, $\hat{r}_{mS}$, $\vec{v}_{mS}$ and $\vec{a}_{mS}$, respectively.

8.4 Implementation of Mach's Principle

We now show how to implement Mach's principle quantitatively based on these postulates and relational forces.

In order to obtain the equation for the conservation of energy and the equation of motion for the test body, we need to include its interaction with all bodies in the universe. We divide these interactions into two groups. (A) The first group defines its interaction with local bodies (the earth, magnets, charges, springs, frictional forces, *etc.*) and with anisotropic distributions of bodies surrounding it (the moon and the sun, the matter around the center of our galaxy, *etc.*) The energy of the test body interacting with all these N bodies will be represented by $U_{Am} = \sum_{j=1}^{N} U_{jm}$, where U_{jm} is the energy of the test body of gravitational mass m_g interacting with body j. The subscript A means anisotropic, but also includes local bodies which may be located around the test body. The force exerted by all these N bodies acting on m_g will be represented by $\vec{F}_{Am} = \sum_{j=1}^{N} \vec{F}_{jm}$, where $\vec{F}_{jm}$ is the force exerted by body j on m_g.

(B) The second group is the interaction of the test body m_g with isotropic distributions of bodies surrounding it. By isotropic distributions we mean bodies scattered with spherical symmetry around m_g such that m_g is inside these distributions, without necessarily being at their center. The energy of interaction of m_g with these isotropic distributions will be represented by U_{Im}, the subscript I meaning isotropic. The force exerted by these isotropic distributions on m_g will be represented by $\vec{F}_{Im}$. We now utilize the known fact that the universe is remarkably isotropic when measured by the integrated microwave and X-ray backgrounds, or by radio source counts and deep galaxy counts [132]. It should be observed that we are not assuming this fact to be true theoretically. We are utilizing this fact as coming from astronomical observations and not as a theoretical hypothesis. Even if one day it is found that the universe is not isotropic in large scale, it will still be possible to derive the main results of relational mechanics. The reason is that even in this case the inertia of the bodies will still be derived from the isotropic part of this anisotropic universe, while its anisotropic part will yield the usual forces.

As the earth does not occupy a central position with respect to the universe, this isotropy on large scale suggests homogeneity on a very large scale. The average density of matter in the universe will not depend on R (the distance of the point under consideration from us): $\rho_g = \rho_o = constant$. Due to the great distance between the galaxies and to their charge neutrality, they can only interact significantly with any distant body through gravitation. We can now integrate Eq. (8.11) utilizing a constant matter density. In this way we obtain U_{Im}, the energy of interaction of m_g with the isotropic part of the distant universe. In other words, with the isotropic and homogeneous distribution of galaxies which is rotating with angular velocity $\vec{\omega}_{US}$ relative to the frame of reference S, shown in Figure 8.3. We replaced $\vec{\omega}_{MS}$ by $\vec{\omega}_{US}$ because we integrated the mass dM of the shell over the whole universe, indicating the mass of the universe by U.

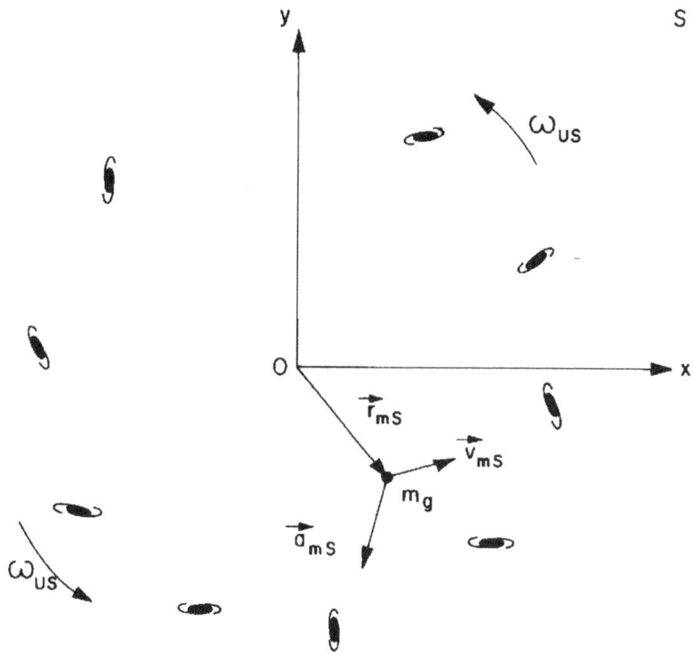

Figure 8.3: Set of distant galaxies rotating relative to S.

$$U_{Im} = -\Phi \left[\frac{3}{\xi} m_g c^2 \right.$$

$$\left. - m_g \frac{(\vec{v}_{mS} - \vec{\omega}_{US} \times \vec{r}_{mS}) \cdot (\vec{v}_{mS} - \vec{\omega}_{US} \times \vec{r}_{mS})}{2} \right] . \tag{8.15}$$

where

$$\Phi \equiv \frac{4\pi}{3} H_g \frac{\xi}{c^2} \int_0^{c/H_o} \rho_g R dR = \frac{2\pi}{3} \xi \frac{H_g \rho_o}{H_o^2} . \tag{8.16}$$

In this last equation, H_o is Hubble's constant and c/H_o is the radius of the known and observable universe.

Analogously, integrating Eq. (8.12) assuming a constant matter density, we also obtain the force $\vec{F}_{Im}$ exerted by this isotropic part of the distant universe on m_g. Once more we are replacing $\vec{\omega}_{MS}$ by $\vec{\omega}_{US}$ as we are integrating the force of the mass dM of the shell over the whole universe, indicating the mass of the universe by U:

$$\vec{F}_{Im} = -\Phi m_g \left[\vec{a}_{mS} + \vec{\omega}_{US} \times (\vec{\omega}_{US} \times \vec{r}_{mS}) \right.$$

$$\left. + 2\vec{v}_{mS} \times \vec{\omega}_{US} + \vec{r}_{mS} \times \frac{d\vec{\omega}_{US}}{dt} \right] . \tag{8.17}$$

By the principle of action and reaction obeyed by Weber's force, we find that body m_g will exert an exactly opposite force on the distant universe.

In a frame of reference S' in which the universe as a whole (the set of distant galaxies) is not spinning, $\vec{\omega}_{US'} = 0$, but in which the universe as a whole moves relative to S' with a velocity $\vec{v}_{US'}$ and translational acceleration $\vec{a}_{US'}$ the integration of Eqs. (8.3) and (8.4) lead to

$$U_{Im} = -\Phi \left[\frac{3}{\xi} m_g c^2 - m_g \frac{(\vec{v}_{mS'} - \vec{v}_{US'}) \cdot (\vec{v}_{mS'} - \vec{v}_{US'})}{2} \right] , \tag{8.18}$$

$$\vec{F}_{Im} = -\Phi m_g (\vec{a}_{mS'} - \vec{a}_{US'}) . \tag{8.19}$$

Here $\vec{v}_{mS'}$ and $\vec{a}_{mS'}$ are the velocity and acceleration of m_g relative to S', shown in Figure 8.4. Again m_g exerts an opposite force on the set of distant galaxies.

If we are in a frame of reference O in which the universe as a whole (the frame of distant galaxies) has no translational or rotational accelerations, $\vec{a}_{UO} = 0$ and $\vec{\omega}_{UO} = 0$, but moves with a constant velocity relative to O, $\vec{v}_{UO}$, Eqs. (8.18) and (8.19) reduce to the simple forms:

$$U_{Im} = -\Phi \left[\frac{3}{\xi} m_g c^2 - m_g \frac{(\vec{v}_{mO} - \vec{v}_{UO}) \cdot (\vec{v}_{mO} - \vec{v}_{UO})}{2} \right] , \tag{8.20}$$

$$\vec{F}_{Im} = -\Phi m_g \vec{a}_{mO} . \tag{8.21}$$

Here $\vec{v}_{mO}$ and $\vec{a}_{mO}$ are the velocity and acceleration of m_g relative to O. Again m_g exerts an opposite force on the set of distant galaxies.

If we are in a frame of reference U in which the universe as a whole (the frame of distant galaxies) is stationary and not rotating, Eqs. (8.18) and (8.19) reduce to the simple forms:

$$U_{Im} = -\Phi \left[\frac{3}{\xi} m_g c^2 - m_g \frac{\vec{v}_{mU} \cdot \vec{v}_{mU}}{2} \right] , \tag{8.22}$$

$$\vec{F}_{Im} = -\Phi m_g \vec{a}_{mU} , \tag{8.23}$$

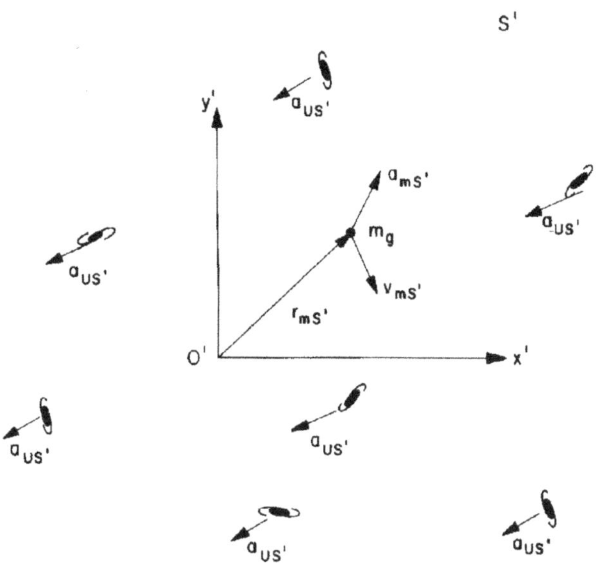

Figure 8.4: Distant universe with a linear acceleration relative to S'.

where $\vec{v}_{mU}$ and $\vec{a}_{mU}$ are the velocity and acceleration of m_g relative to the frame of reference U.

We are now in a position to implement Mach's principle quantitatively utilizing our third postulate that the sum of all forces is zero (or that the sum of the interaction energies is zero), namely:

$$U_{Am} + U_{Im} = 0 \ , \tag{8.24}$$

$$\vec{F}_{Am} + \vec{F}_{Im} = 0 \ . \tag{8.25}$$

In the frame of reference S this yields:

$$U_{Am} + U_{Im} = \sum_{j=1}^{N} U_{jm} - \Phi \left[\frac{3}{\xi} m_g c^2 \right.$$

$$\left. - m_g \frac{(\vec{v}_{mS} - \vec{\omega}_{US} \times \vec{r}_{mS}) \cdot (\vec{v}_{mS} - \vec{\omega}_{US} \times \vec{r}_{mS})}{2} \right] = 0 \ , \tag{8.26}$$

$$\vec{F}_{Am} + \vec{F}_{Im} = \sum_{j=1}^{N} \vec{F}_{jm} - \Phi m_g \left[\vec{a}_{mS} + \vec{\omega}_{US} \times (\vec{\omega}_{US} \times \vec{r}_{mS}) \right.$$

$$+ 2\vec{v}_{mS} \times \vec{\omega}_{US} + \vec{r}_{mS} \times \frac{d\vec{\omega}_{US}}{dt}\Bigg] = 0 \; . \tag{8.27}$$

In the frame of reference S' we have:

$$U_{Am} + U_{Im} = \sum_{j=1}^{N} U_{jm} - \Phi\left[\frac{3}{\xi}m_g c^2 - m_g \frac{(\vec{v}_{mS'} - \vec{v}_{US'}) \cdot (\vec{v}_{mS'} - \vec{v}_{US'})}{2}\right] = 0 \; ,$$
$$\tag{8.28}$$

$$\vec{F}_{Am} + \vec{F}_{Im} = \sum_{j=1}^{N} \vec{F}_{jm} - \Phi m_g(\vec{a}_{mS'} - \vec{a}_{US'}) = 0 \; . \tag{8.29}$$

In the frame of reference O we have:

$$U_{Am} + U_{Im} = \sum_{j=1}^{N} U_{jm} - \Phi\left[\frac{3}{\xi}m_g c^2 - m_g \frac{(\vec{v}_{mO} - \vec{v}_{UO}) \cdot (\vec{v}_{mO} - \vec{v}_{UO})}{2}\right] = 0 \; ,$$
$$\tag{8.30}$$

$$\vec{F}_{Am} + \vec{F}_{Im} = \sum_{j=1}^{N} \vec{F}_{jm} - \Phi m_g \vec{a}_{mO} = 0 \; . \tag{8.31}$$

Finally, in the frame of reference U we have:

$$U_{Am} + U_{Im} = \sum_{j=1}^{N} U_{jm} + -\Phi\left[\frac{3}{\xi}m_g c^2 - \frac{m_g \vec{v}_{mU} \cdot \vec{v}_{mU}}{2}\right] = 0 \; , \tag{8.32}$$

$$\vec{F}_{Am} + \vec{F}_{Im} = \sum_{j=1}^{N} \vec{F}_{jm} - \Phi m_g \vec{a}_{mU} = 0 \; . \tag{8.33}$$

Here $\vec{v}_{mU}$ and $\vec{a}_{mU}$ are the velocity and acceleration of m_g relative to the frame of distant galaxies. In this frame the set of distant galaxies is seen at rest without any linear or angular velocity, and without any linear or angular acceleration. We will call this the universal frame of reference U, as in Figure 8.5.

Contrary to Newton's absolute space which was "without relation to anything external," this universal frame of reference is completely determined by the external material world. It is the frame in which the distant matter as a whole is at rest, despite the peculiar velocities of the galaxies in this frame. In

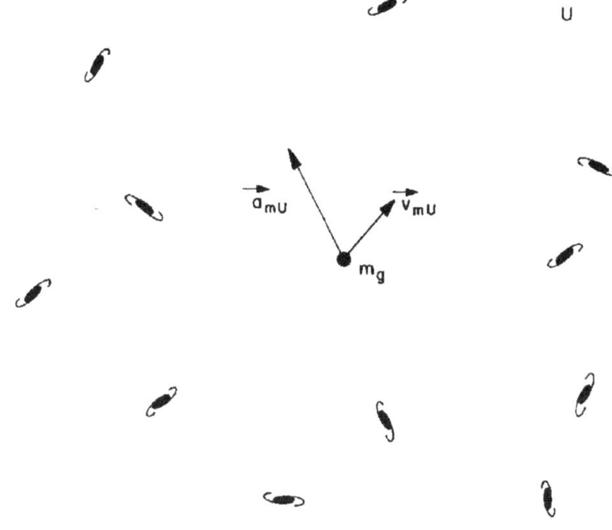

Figure 8.5: Universal frame of reference U.

this universal frame the universe appears isotropic in the large, with distant galaxies appearing uniformly distributed. This is presumably also the same frame in which the cosmic background radiation is seen as isotropic without the dipole anisotropy. It is in this frame that the equation of motion of relational mechanics takes the simplest form, without the appearance of terms containing the acceleration of the distant universe.

If we had utilized Eqs. (8.5) and (8.6), we could have integrated R from zero to infinity, without divergences. Then we would have obtained equations (8.15) to (8.33) with $A \equiv 4\pi\xi H_g\rho_o/3H_o^2$ instead of Φ, if in (8.5) and (8.6) we had utilized $\alpha = H_o/c$ ([12], Chapt. 7):

$$A \equiv \frac{4\pi}{3}H_g\frac{\xi}{c^2}\rho_o \int_0^\infty Re^{-H_oR/c}dR = \frac{4\pi}{3}\xi\frac{H_g\rho_o}{H_o^2} \ . \tag{8.34}$$

It should be emphasized that the ρ_o which appears in Eqs. (8.16) and (8.34) is the volumetric matter density of galaxies in space (N times the average mass of each galaxy, divided by the volume occupied by these N galaxies, with $N \gg 1$). Moreover, we are integrating over the whole known universe. This means that the main contribution to Φ or to A will come from external and distant galaxies, and not from the stars belonging to our own Milky Way galaxy. We will see this in more details in the next section.

This completes the mathematical implementation of Mach's principle. We now discuss all the direct consequences we can obtain from relational mechanics.

8.5   General Consequences

Eq. (8.32) is similar to the classical equation for the conservation of energy in an inertial frame of reference. Eq. (8.33) is similar to Newton's second law of motion in an inertial frame of reference. In order to see this in more detail we consider two bodies 1 and 2 interacting gravitationally with one another and with the distant universe. Supposing $\dot{r}_{12}^2 \ll c^2$, $|r_{12}\ddot{r}_{12}| \ll c^2$ and utilizing Eqs. (8.3), (8.4), (8.32) and (8.33) in the universal frame of reference yields the following equations for body 1:

$$-\frac{H_g}{\Phi}\frac{m_{g1}m_{g2}}{r_{12}} + \frac{m_{g1}v_{1U}^2}{2} = \frac{3}{\xi}m_{g1}c^2 \ , \tag{8.35}$$

$$-\frac{H_g}{\Phi}m_{g1}m_{g2}\frac{\hat{r}_{12}}{r_{12}^2} = m_{g1}\vec{a}_{1U} \ . \tag{8.36}$$

This last equation will be equivalent to Newton's second law of motion in this case if $H_g/\Phi = 3H_o^2/2\pi\xi\rho_o$ becomes the Newtonian constant of universal gravitation $G = 6.67 \times 10^{-11}$ Nm^2/kg^2:

$$\frac{H_g}{\Phi} = G \leftrightarrow \frac{3H_o^2}{2\pi\xi\rho_o} = G \ . \tag{8.37}$$

This remarkable relation connecting three independent and measurable (or observable) magnitudes of physics (G, H_o and ρ_o) is a necessary consequence of any model that seeks to implement Mach's principle. This relation has been known to be approximately true (with ξ between 1 and 20) since the 1930's with Dirac's large numbers [133]. But while for Dirac this was derived more like numerology without a deeper understanding, here it is derived from first principles as a consequence of relational mechanics. We also know that this relation must be true from the validity of Newtonian mechanics in everyday laboratory experiments. In other words, as we recover the Newtonian results only with $H_g/\Phi = G$, we conclude that this must be the case. But the more remarkable fact is that this can be obtained independently utilizing the known values of G, H_o and ρ_o. The value of G is 6.67×10^{-11} Nm^2/kg^2, while $\rho_o/H_o^2 \approx 4.5 \times 10^8$ kgs^2/m^3 [134, Sections 2.2 and 2.3, pp. 44-74]. The greatest uncertainty is in the value of ρ_o/H_o^2, which is not yet accurately known.

If instead of integrating up to Hubble's radius we had integrated only up to the radius of our own flat galaxy, $R_s \approx \times 10^{20}$ m, with a total mass $M_s \approx 4 \times 10^{41}$ kg, supposing an average matter density $\rho_s \approx 10^{-21}$ kg/m^3, Eq. (8.16) would become:

$$\Phi_{star} \equiv \frac{4\pi}{3} H_g \frac{\xi}{c^2} \int_0^{R_s} \rho_g R dR = \frac{2\pi}{3} \xi \frac{H_g \rho_s R_s^2}{c^2} \ . \tag{8.38}$$

But this would not allow the mathematical implementation of Mach's principle based on the principle of dynamical equilibrium. The reason is that Φ/Φ_{star} is given approximately by:

$$\frac{\Phi}{\Phi_{star}} = \frac{\rho_o c^2}{H_o^2 \rho_s R_s^2} \approx 10^7 \ . \tag{8.39}$$

This means that the contribution of the fixed stars belonging to our own galaxy to the inertia of any body is negligible compared with the contribution of the distant galaxies.

If we had chosen Eqs. (8.5), (8.6), with $\alpha = H_o/c$, and had integrated to infinity we would arrive at $H_g/A = G$ instead of $H_g/\Phi = G$:

$$\frac{H_g}{A} = G \leftrightarrow \frac{3H_o^2}{4\pi\xi\rho_o} = G \ . \tag{8.40}$$

Equations (8.37) and (8.40) show that the value of H_g is not determined in relational mechanics.

Analogously, suppose we have two point charges q_1 and q_2 interacting with one another and with the isotropic distribution of galaxies around them. We will also consider $\dot{r}_{12}^2 \ll c^2$, $|r_{12}\ddot{r}_{12}| \ll c^2$. The equations of motion for charge 1 can be obtained in the universal frame of reference from Eqs. (8.1), (8.2), (8.32) and (8.33):

$$\frac{H_e}{\Phi} \frac{q_1 q_2}{r_{12}} + \frac{m_{g1} v_{1U}^2}{2} = \frac{3}{\xi} m_{g1} c^2 \ , \tag{8.41}$$

$$\frac{H_e}{\Phi} q_1 q_2 \frac{\hat{r}_{12}}{r_{12}^2} = m_{g1} \vec{a}_{1U} \ . \tag{8.42}$$

This last equation is analogous to the equation of motion describing the interaction of two point charges in Newtonian mechanics provided

$$\frac{H_e}{\Phi} = \frac{1}{4\pi\varepsilon_o} \leftrightarrow \frac{3H_o^2}{2\pi\xi\rho_o} \frac{H_e}{H_g} = \frac{1}{4\pi\varepsilon_o} \ . \tag{8.43}$$

Analogously, if we had utilized the exponential decay and integrated until infinity we would obtain:

$$\frac{H_e}{A} = \frac{1}{4\pi\varepsilon_o} \leftrightarrow \frac{3H_o^2}{4\pi\xi\rho_o} \frac{H_e}{H_g} = \frac{1}{4\pi\varepsilon_o} \ . \tag{8.44}$$

Although the value of H_g is not determined, we can obtain the ratio H_e/H_g.

The same thing happens with the other forces. For instance, the elastic force and the frictional force proportional to the velocity are written in relational mechanics as $-K(\vec{\ell} - \vec{\ell}_o)$ and $-B\vec{v}$, respectively. Here K and B are constants, $\vec{\ell} - \vec{\ell}_o$ is the displacement of the spring from the equilibrium distance ℓ_o and $\vec{v}$ is the velocity of the test body relative to the medium (air or water, for instance) with which it is interacting. From Eq. (8.33) we arrive in these two cases (test body interacting with a spring and the distant universe, or with a resistive medium and the distant universe) at: $-(K/\Phi)(\vec{\ell} - \vec{\ell}_o) = m_g\vec{a}_{mU}$ and $-(B/\Phi)\vec{v} = m_g\vec{a}_{mU}$. These equations will be analogous to the similar ones in classical mechanics if

$$\frac{K}{\Phi} = k , \qquad (8.45)$$

and

$$\frac{B}{\Phi} = b , \qquad (8.46)$$

where k and b are the elastic constant and frictional constants of classical mechanics, respectively. We will assume this is valid for these cases and for other forces.

In relational mechanics we derive an inertia analogous to Newtonian inertia from a generalized law of gravitation (Weber's law). The opposite approach of deriving gravitation from inertia has been taken by Roscoe [135].

Eq. (8.32) is analogous to the equation of conservation of energy in classical mechanics in inertial frames of reference, while Eq. (8.33) is analogous to Newton's second law of motion in absolute space or in inertial frames of reference. But the difference is that now we have *derived* an expression analogous to the kinetic energy and another expression analogous to the Newtonian equation of motion. In classical mechanics, we were obliged to begin with the concept of kinetic energy (without knowing where it came from). Likewise, Newton was obliged to begin with $\vec{F} = d(m_i\vec{v})/dt$ or $\vec{F} = m_i\vec{a}$, as he could not derive it. For this reason, the concept of inertia or inertial mass had to be introduced *a priori*.

In relational mechanics, we are deriving an energy analogous to the classical kinetic energy. But when we identify the $m_g v_{mU}^2/2$ of Eq. (8.32) with the classical kinetic energy $m_i v^2/2$, we at once understand the mysterious proportionality between inertial and gravitational masses which had appeared in Newtonian mechanics. In other words, in relational mechanics we found that the kinetic energy is an interaction energy, like any other kind of potential energy. It is an energy of gravitational origin arising from the relative motion between m_g and the universe as a whole surrounding it. It is no longer frame-dependent, as was the case with the classical kinetic energy. The reason is that the energy U_{Im} has the same numerical value (although not necessarily the same form) in all frames of reference. For instance, in the frame O in which the universe as a whole (the frame of distant galaxies) is not rotating but is translating with velocity $\vec{v}_{UO}$, the kinetic energy of relational mechanics is be given by $m_g|\vec{v}_{mO} - \vec{v}_{UO}|^2/2$ instead of $m_g v_{mU}^2/2$. Obviously $m_g v_{mU}^2/2 = m_g|\vec{v}_{mO} - \vec{v}_{UO}|^2/2$ because $\vec{v}_{mO} = \vec{v}_{mU} + \vec{v}_{UO}$.

The same can be concluded from Eq. (8.33). Identifying this equation of relational mechanics with Newton's second law of motion, yields at once the proportionality between inertial and gravitational masses. The force $-m_g\vec{a}_{mU}$ of Eq. (8.33) is a real gravitational force between m_g and the universe at large (the set of distant galaxies) when there is a relative acceleration between them. The force $\vec{F}_{Im}$ is not frame-dependent. It has the same numerical value and points in the same direction relative to other matter in all frames of reference, although it does not necessarily have the same form. For instance, in the universal frame U we have: $\vec{F}_{Im} = -m_g\vec{a}_{mU}$. In another frame S which rotates relative to the frame of distant galaxies, we have: $\vec{F}_{Im} = -m_g[\vec{a}_{mS} + \vec{\omega}_{US} \times (\vec{\omega}_{US} \times \vec{r}_{mS}) + 2\vec{v}_{mS} \times \omega_{US} + \vec{r}_{mS} \times d\vec{\omega}_{US}/dt]$. In another frame S' which does not rotate relative to the distant galaxies but which has a translational acceleration relative to the distant matter, we have: $\vec{F}_{Im} = -m_g(\vec{a}_{mS'} - \vec{a}_{US'})$. Although the form of $\vec{F}_{Im}$ is different in these three frames of reference U, S and S', the numerical value is the same. For this reason, we did not need to specify $\vec{F}_{Im}^U$, $\vec{F}_{Im}^S$ or $\vec{F}_{Im}^{S'}$. Not only is the numerical value the same, but the direction of $\vec{F}_{Im}$ is the same in all these frames, always pointing to the same point relative to other matter. If this force $\vec{F}_{Im}$ on m_g is directed at a specific time to one galaxy, such as Andromeda, it will point to this galaxy in all frames of reference. For instance, suppose we have two bodies 1 and 2 connected by a spring and oscillating along the line of their junction in the frame U, and that the line of their junction is the same line connecting the center of our galaxy to the center of Andromeda. Then the forces $\vec{F}_{I1}$ and $\vec{F}_{I2}$ will also point along the line connecting our galaxy to Andromeda when calculated in all frames of reference. This is due to the fact that Weber's force depends only on relational quantities like r_{12}, $\dot{r}_{12}$ and $\ddot{r}_{12}$, which have the same value in all frames of reference.

Suppose now the case in which the resultant anisotropic force $\vec{F}_{Am} = \sum_{j=1}^{N} \vec{F}_{jm}$ acting on m_g is zero. From Eq. (8.33) and from $m_g \neq 0$ we obtain that $\vec{a}_{mU} = 0$. In other words, we conclude that the test body will move with a constant velocity relative to the frame in which the set of distant galaxies is at rest. Identifying relational mechanics with Newtonian mechanics shows that we have *derived* a law similar to Newton's first law of motion. But now, instead of saying that a body will move with a constant velocity relative to absolute space (an entity to which we have no access), we say that the body will move with constant velocity relative to the frame of distant galaxies. If this is the case, the test body will also move with a constant velocity relative to any other frame O which moves with a constant velocity relative to the frame of distant galaxies. These reference frames may then be identified with the inertial frames of classical mechanics. But now they have been completely determined by distant matter.

Identification of Eqs. (8.32) and (8.33) with the classical equation for the conservation of energy and with Newton's second law of motion explains the proportionality between inertial and gravitational masses of Newtonian mechanics. The concepts of inertia of a body, of inertial mass, of inertial frames of reference, of kinetic energy, *etc.* have not been introduced in relational mechanics. It is only when we identify Eqs. (8.32) and (8.33) with the analogous equations of classical mechanics that we understand and solve this puzzle of Newtonian theory. We explain why the Newtonian inertial and gravitational masses are proportional to one another. The reason is that the second terms on the left of Eqs. (8.32) and (8.33) arose from gravitational interactions between the gravitational mass m_g and the gravitational mass of the distant galaxies when there is a relative motion between m_g and these galaxies. The mass of $m_g v_{mU}^2/2$ or of $m_g \vec{a}_{mU}$ is the gravitational mass of the test body. Only when we identify these terms with Newtonian mechanics, where we have $m_i v^2/2$ and $m_i \vec{a}$, does it become clear that these "kinetic" expressions of Newtonian mechanics have a gravitational origin. Newtonian mechanics gains a new meaning and a clear understanding when viewed from the standpoint of relational mechanics.

In relational mechanics we do not need to postulate the proportionality or equality between m_g and m_i, as must be done in Einstein's general theory of relativity. Einstein had postulated the equality between m_g and m_i in his principle of equivalence. As he postulated this relation without supplying more fundamental explanations for its origin, we can see that he could not derive it. On the other hand, in relational mechanics this result is a direct consequence of the theory. This shows the great advantage of relational mechanics over Einstein's general theory of relativity.

Another thing which is explained at once in relational mechanics is the equality between $\vec{\omega}_k$ and $\vec{\omega}_d$, the kinematical and dynamical rotations of the earth. As we have just seen, the equations of motion only take the simple form of Eqs. (8.32) and (8.33) in a frame of reference relative to which the universe as a whole (the set of distant galaxies) is stationary. In another frame of reference S there must appear terms in the energy U_{Im} and in the force $\vec{F}_{Im}$ which depend on $\vec{v}_{US}$, on $\vec{a}_{US}$ and on $\vec{\omega}_{US}$ (the velocity, translational acceleration and rotational velocities of the universe as a whole relative to S). When we identify this fact and relational mechanics with Newtonian mechanics everything becomes obvious and understandable. The explanation for the coincidence of Newtonian mechanics ($\vec{\omega}_d = \vec{\omega}_d$) is that the distant universe as a whole defines the best inertial frame (which is the frame in which Newton's laws of motion are valid without the introduction of centrifugal and Coriolis forces). This means that the set of distant galaxies does not rotate relative to absolute space (this is the meaning of the observational fact $\vec{\omega}_k = \vec{\omega}_d$) because it defines what absolute space is.

In a frame of reference S in which the universe as a whole is spinning relative to the center of this frame with $\vec{\omega}_{US}(t)$, but without translational acceleration, Eq. (8.27) takes the form

$$\sum_{j=1}^{N} \vec{F}_{jm} - \Phi m_g \left[\vec{a}_{mS} + \vec{\omega}_{US} \times (\vec{\omega}_{US} \times \vec{r}_{mS}) \right.$$

$$\left. + 2\vec{v}_{mS} \times \vec{\omega}_{US} + \vec{r}_{mS} \times \frac{d\vec{\omega}_{US}}{dt} \right] = 0 \ . \tag{8.47}$$

Here $\vec{r}_{mS}$ is the position vector of m_g relative to the origin of S. Moreover, $\vec{v}_{mS}$ and $\vec{a}_{mS}$ are the velocity and acceleration of the test body relative to this frame of reference S.

This result of relational mechanics has the same form as Newton's second law of motion with fictitious forces. Identifying these two formulae leads to the conclusion that the centrifugal and Coriolis forces are not fictitious forces anymore. Quite the contrary, in relational mechanics they are seen as real forces of gravitational origin arising from the interaction of the accelerated test body and the spinning universe around it. This is in almost complete agreement with Mach's ideas, as we have shown that 'when the heaven of galaxies is rotated, centrifugal forces arise!' The only difference is that Mach knew only the existence of the set of fixed stars. It was in 1924 that Hubble established the existence of external galaxies. This came after Mach's death in 1916. Here we have shown that rotating only our own galaxy (*i.e.*, the set of fixed stars) relative to an observer does not yield enough sensible centrifugal force. On the other hand, the rotation of the whole known universe (the set of distant galaxies) will yield exactly the full centrifugal force observed to exist in frames relative to which the set of galaxies is rotating.

Let us now discuss how to derive another aspect which was correctly pointed out by Mach. In Eq. (8.33) the acceleration which appears is the acceleration of the test body m_g relative to the frame of distant galaxies. It is the acceleration of the test body in a reference frame relative to which the set of galaxies as a whole does not rotate and has no translational acceleration. When we are dealing with bodies on the surface of the earth, performing collision experiments with billiard balls, for instance, we usually measure only the accelerations of the test bodies relative to the earth. However, the earth rotates around its axis relative to the frame of fixed stars with a period of one day, and translates around the sun relative to the frame of fixed stars with a period of one year, while the solar system translates around the center of our galaxy relative to the frame of distant galaxies with a period of approximately 2.5×10^8 years. As these are all accelerated motions relative to the distant universe, all of them should be taken into account when we try to apply Eq. (8.33) to study the motion of the test body. However, there is a situation in which these aspects are greatly simplified. Suppose that this acceleration of the test body relative to the earth is much greater than the acceleration of its location on the earth's surface relative to the frame of distant galaxies. There are three main components in the acceleration of the earth at each point of its surface relative to the frame of distant galaxies. The first is the centripetal acceleration due to the rotation of the earth every day relative to the distant stars, which is given at the equator by (v_t being the tangential velocity and R the radius of curvature): $a_r \approx v_t^2/R \approx (4.6 \times 10^2 \text{ m/s})^2/(6.4 \times 10^6 \text{ m}) \approx 3.4 \times 10^{-2} \text{ m/s}^2$. The second is the centripetal acceleration due to the translation of the earth around the sun every year relative to the distant stars, which is given roughly by: $a_t \approx (3 \times 10^4 \text{ m/s})^2/(1.5 \times 10^{11} \text{ m}) \approx 6 \times 10^{-3} \text{ m/s}^2$. The third is the centripetal acceleration of the solar system due to its orbit around the center of our galaxy relative to the frame of the distant galaxies every 2.5×10^8 years, which is given approximately by: $a_s \approx (2 \times 10^5 \text{ m/s})^2 \: / \: (2.5 \times 10^{20} \text{ m}) \approx 10^{-10}$ m/s^2. Calling the acceleration of the test body m_g relative to the earth $\vec{a}_{me}$ there is a situation in which:

$$a_{me} \gg a_r > a_t \gg a_s \; . \tag{8.48}$$

In this case, we can disregard the components of the body's acceleration related to the motion of the earth relative to the frame of distant galaxies, and consider only the acceleration of the test body relative to the earth. When this condition is satisfied we can say that during this experiment the set of distant galaxies will be essentially without acceleration relative to the frame of fixed stars, and also relative to the earth. This means that accelerations relative to distant galaxies can be conveniently described with good accuracy by accelerations relative to the earth. Eq. (8.33) can then be approximated whenever Eq. (8.48) is satisfied in the simple form given by

$$\sum_{j=1}^{N} \vec{F}_{jm} - \Phi m_g \vec{a}_{me} = 0 \; . \tag{8.49}$$

For experiments over the earth's surface in which the test body moves with accelerations satisfying Eq. (8.48) we can consider to a good approximation only its acceleration relative to the earth's surface.

However, if we are studying the acceleration of the earth as a whole in its orbit around the sun, or a test body moving over the earth's surface with a very small acceleration of the order of $a_r \approx 3.4 \times 10^{-2}$ m/s^2 or of $a_t \approx 6 \times 10^{-3}$ m/s^2, then we also need to take these accelerations into account. In these cases, and due to the fact that the centripetal acceleration of the solar system, $\approx 10^{-10}$ m/s^2, is much smaller than a_r or a_t, we can write Eq. (8.33) in the form

$$\sum_{j=1}^{N} \vec{F}_{jm} - \Phi m_g \vec{a}_{mf} = 0 \ . \tag{8.50}$$

In this equation, $\vec{a}_{mf}$ is the acceleration of the test body relative to the frame of fixed stars (frame in which the set of fixed stars is seen without rotation and without translational acceleration). This equation in this form should be applied in the frame of fixed stars when the following condition is satisfied:

$$a_{mf} \approx a_r > a_t \gg a_s \ . \tag{8.51}$$

In this case we may say that the set of distant galaxies is essentially without acceleration relative to the frame of fixed stars. It is then possible and convenient to refer motions to the fixed stars instead of referring them to the distant galaxies.

In the case in which $\sum_{j=1}^{N} \vec{F}_{jm} = 0$, we find from Eq. (8.49) that the test body will move relative to the earth's surface with a constant velocity. If we need a better approximation, then we find that in this case of zero anisotropic resultant force Eq. (8.50) leads to the conclusion that the test body will move with a constant velocity relative to the frame of fixed stars. An even better approximation utilizing Eq. (8.33) leads to the conclusion that the test body will move with a constant velocity relative to the frame of distant galaxies.

The meaning of Mach's statement that "I have remained to the present day the only one who insists upon referring the law of inertia to the earth, and in the case of motions of great spatial and temporal extent, to the fixed stars" is now clear. Obviously, in his time the existence of external galaxies was not yet known, but it is essentially the same thing. The important fact obtained with relational mechanics is that the law of inertia has now been derived as meaningful when motion is referred to material bodies. These material bodies can be the earth, the fixed stars or the distant galaxies.

In relational mechanics there are only relational quantities. That is, only velocities and accelerations of the test body relative to the other bodies in the universe with which it is interacting are meaningful. For this reason it is not necessary to worry about coordinate transformations (Galilean transformation, Lorentz's transformation, *etc.*). This is one of the main advantages of relational mechanics when compared with classical physics.

We now summarize the main direct consequences of relational mechanics when we identify it with Newtonian mechanics:

A) We *derive* equations similar to Newton's first and second laws of motion.

B) We *derive* the proportionality between inertial and gravitational masses.

C) We *derive* the fact that the best inertial frame we have is the frame of distant galaxies. In other words, we derive the observed fact that $\vec{\omega}_k = \vec{\omega}_d$.

D) We *derive* the kinetic energy as an interaction energy of gravitational origin between the test body and the distant galaxies.

E) We *derive* the fact that all fictitious forces of Newtonian mechanics are in fact real forces of gravitational origin acting between the test body and the distant galaxies.

F) We *derive* the relation known to be true between G, H_o and ρ_o, namely, $3H_o^2 = 2 - 4\pi\xi G\rho_o$, with $\xi = 6$ as we will see later.

G) We *derive* that the "inertial" forces $\vec{F}_{Im}$ have the same numerical value in all frames of reference, though not necessarily the same form.

8.6 Cosmology

We now discuss Eqs. (8.37) and (8.40). They can also be written as $R_o^2 = 3c^2/2\pi\xi G\rho_o$ or $R_o^2 = 3c^2/4\pi\xi G\rho_o$, where $R_o = c/H_o$ is the radius of the known universe. If the universe is expanding, R_o will be a function of time. This means that $c^2/\xi G\rho_o$ will also be a function of time.

But the idea of the expansion of the universe and the related big bang arose from the assumption that Hubble's law of redshifts ($\triangle\lambda/\lambda_o \approx rH_o/c$) is due to a Doppler effect caused by the recession between galaxies. Our own point of view, however, is that the cosmological redshift (related to Hubble's law) is due to some kind of effect known in the literature as "tired-light" and not to a Doppler effect. We believe Hubble's law is due to an interaction between the light emitted by a distant galaxy and the intergalactic matter. It seems to us that the principal cause of this redshift is the loss of photon energy as it interacts with the intergalactic medium. With this supposition Hubble's law can be easily derived without assuming it to be due to a Doppler effect [136]. Essentially we utilize the photon energy as $E = h\nu = hc/\lambda$, where $h = 6.6 \times 10^{-34}$ Js is Planck's constant. We also assume as usual an exponential decay for the energy due to its interaction with the intervening medium, namely: $E(r) = E_o e^{-H_o r/c}$, where E_o is the emitted energy in the distant galaxy at a distance r from our own and $E(r)$ is the energy arriving here. From these two expressions the redshift $z \equiv (\lambda(r) - \lambda_o)/\lambda_o$ is found to be given by $e^{H_o r/c} - 1 \approx H_o r/c$. This expresssion has been derived by many writers since 1929. No Doppler effect has been assumed in this derivation.

We are not yet sure what kind of mechanism is at work here (photon-photon interaction, inelastic collision between photons and free electrons, or between photons and molecules, *etc.*). Nevertheless, we have explored this possibility in other works ([35], [136], [36], [137] and [138]). In these articles many more references can be found to other authors working along the same lines. In essential aspects we are continuing the works of Regener, Walther Nernst, Finlay-Freundlich, Max Born and Louis de Broglie on an equilibrium cosmology without expansion (see Regener [139] with an English translation in [140], Nernst [141] and [142] with English translations in [143] and [144], Finlay-Freundlich [145], [146], [147], Born [148], [149] and de Broglie [150]). As there is no expansion of the universe in this model, it does not need a continuous creation of matter, as required by the steady-state model of Hoyle, Bondi and Gold. Our model is a universe in dynamical equilibrium without expansion and without creation of matter. It should be emphasized here that Walther Nernst (the father of the third law of thermodynamics and Nobel prize winner) and Louis de Broglie (one of the founders of quantum mechanics and Nobel prize winner) never accepted the idea of the big bang, always working with a model of the universe in dynamical equilibrium without expansion. For more recent developments and different approaches to these models of a universe in dynamical equilibrium without expansion, see [151] and papers therein, [152], [153], [154], [155], [156], [157], [158], [159], [160], [161], [162], [163], [164], [165], [166], [167], [162], [168], [169], [96] *etc.*

Hubble himself had doubts that the cosmological redshift was due to a Doppler effect. He suggested it might be due to a new principle of nature [170, pp. 30, 63 and 66], [171] and [172, pp. 88-89, 121-123, 193 and 197].

It is important to discuss here the cosmic background radiation, CBR. This is a radiation with the spectrum characteristic of a black body having a temperature of 2.7 K. Usually it is claimed that the CBR is a proof of the big bang and of the expansion of the universe as it had been predicted by Gamow and collaborators (proponents of the big bang) prior to its discovery by Penzias and Wilson in 1965 [26]. However, we performed a bibliographic search and found something quite different from this view [136], [36], [173], [138]. The main point to be stressed here is that the published predictions of this temperature made by Gamow and collaborators (based on the big bang) were of 5 K in 1948, > 5 K in 1949, 7 K in 1953 and 50 K in 1961 [174], [175], [147] and [176, pp. 42-43]. The values were always increasing and each time they departed more and more from the 2.7 K found later on! On the other hand we have found several predictions or estimations of this temperature based on a stationary universe without expansion, always varying between 2 K and 6 K. Moreover, one of these estimates was performed in 1896, prior to Gamow's birth in 1904! The estimates are: 5 K < T < 6 K (Guillaume in 1896); 3.1 K (Eddington in 1926); 2.8 K (Regener in 1933 and Nernst in 1937 and 1938) and 1.9 K < T <

6.0 K (Finlay-Freundlich and Max Born between 1953 and 1954): Guillaume [177] with partial English translation in [173], Eddington [178], Regener [139] with English translation in [140], Nernst [141] with English translation in [143], [142] with English translation in [144], Herzberg [179, p. 496], Finlay-Freundlich [145], [146], [147], Born [148] and [149].

The conclusion is that the discovery of the CBR by Penzias and Wilson in 1965 is a decisive factor in favour of a universe in dynamical equilibrium without expansion, and against the big bang.

Our own cosmological model is an eternal universe (not being created) without boundaries (extending indefinitely in all directions). For this reason we prefer to integrate to infinity and utilize Eqs. (8.34) and (8.40), instead of (8.16) and (8.37). In this case $R_o = c/H_o$ would be seen as a characteristic length of gravitational interactions, instead of denoting the radius or size of the universe. This means that for us, R_o and all other quantities (like c, ξ, and ρ_o) are constants, and not a function of time. As up to now we have written all equations of motion in relational mechanics in terms of Φ and not A, we will continue doing so here to avoid repetition.

8.7   Ptolemaic and Copernican World Views

As we have seen, Leibniz and Mach emphasized that the Ptolemaic geocentric system and the Copernican heliocentric system are equally valid and correct. With relational mechanics we implement this quantitatively, showing the equivalence of both world views.

Let us consider motions over the earth's surface and in the solar system such that we can neglect the acceleration of the solar system relative to the frame of distant galaxies (with a typical value of 10^{-10} m/s^2). Moreover, as the mass of the sun is much greater than the mass of the planets, we can, in a first approximation, disregard the motion of the sun relative to the fixed stars due to the gravitational attraction of the other planets, as compared with the motion of the planets relative to the stars. We can then say that the sun is essentially at rest relative to the fixed stars, while the earth and other planets move relative to them.

We first consider the Copernican world view, which is usually seen as being proved to be true by Galileo and Newton. Here we consider the sun in the center of the universe while the earth and the planets orbit around it and rotate around their axes relative to the frame of fixed stars. To simplify the analysis we consider only circular orbits. Relational mechanics can be applied here with astounding success in the form of Eq. (8.50). Despite the gravitational attraction between the sun and the planets, the earth and other planets do not fall into the sun because they have an acceleration relative to the fixed stars.

The distant matter in the universe exerts a force $-m_g\vec{a}_{mf}$ on accelerated planets, keeping them in their annual orbits. The rotation of the earth around its axis relative to the fixed stars explains its oblate form, with a smaller distance between the poles than at the equator between east and west. Foucault's pendulum is explained by noting that the plane of oscillation remains fixed relative to the fixed stars *etc.*

In the Ptolemaic system the earth is considered to be at rest and without rotation in the center of the universe, while the sun, other planets and fixed stars rotate around the earth. In relational mechanics this rotation of distant matter yields the force (8.17) such that the equation of motion takes the form of Eq. (8.47). Now the gravitational attraction of the sun is balanced by a real gravitational centrifugal force due to the annual rotation of distant masses around the earth (with a component having a period of one year). In this way the earth can remain at rest and at an essentially constant distance from the sun. The diurnal rotation of distant masses around the earth (with a period of one day) yields a real gravitational centrifugal force flattening the earth at the poles. Foucault's pendulum is explained by a real Coriolis force acting on moving masses over the earth's surface in the form $-2m_g\vec{v}_{me} \times \vec{\omega}_{Ue}$, where $\vec{v}_{me}$ is the velocity of the test body relative to the earth and $\vec{\omega}_{Ue}$ is the angular rotation of the distant masses around the earth. The effect of this force will be to keep the plane of oscillation of the pendulum rotating together with the fixed stars, *etc.*

As a matter of fact, any other frame of reference would be equally valid. Anyone or any arbitrary frame of reference can be considered at rest, while the entire universe moves relative to this person according to his will. This happens not only kinematically as has been always known, but also dynamically. All local forces acting on the person will be balanced by the force exerted by the rest of the universe in such a way that the acceleration and velocity of the person will be always zero. For instance, consider a rock falling freely to the earth due to its weight $\vec{P}$. In the frame of the rock it will always remain at rest, while the earth and the universe are accelerated upwards (in the direction of the earth towards the rock). This happens in such a way that the gravitational force $\vec{P}$ exerted by the earth on the rock is balanced by a gravitational force exerted by the distant masses on the rock with a value $m_g\vec{a}_{Um}$ such that $\vec{P} + m_g\vec{a}_{Um} = 0$, where $\vec{a}_{Um}$ is the acceleration of distant masses relative to the rock of gravitational mass m_g.

Relational mechanics implements quantitatively and dynamically the old objective of making all frames of reference equally valid and correct. The form of the force exerted by distance masses on a test body may be different in different frames of reference, but not the value or direction of this force relative to other masses. It may be more practical, simple or mathematically convenient to consider one frame of reference as preferred relative to others, but as a matter of fact all of them will yield the same dynamical consequences (although it may be more difficult to perform the calculations in certain frames in order to arrive at the final results).

An important consequence of relational mechanics is that kinematically equivalent motions have been shown to be dynamically equivalent. Regardless of whether we say that the stars and galaxies are at rest while the earth rotates around its axis with a period of one day, or that the earth is at rest while the stars and galaxies rotate in the opposite direction relative to the earth with a period of one day, in both cases relational mechanics will yield the flattening of the earth as a consequence of this relative rotation. No other theory of mechanics ever proposed has implemented quantitatively this consequence. Although other theorists have tried to implement this philosophical and aesthetic appealing consequence, no one has ever succeeded. What was missing was a relational force law like Weber's.

8.8   Implementation of Einstein's Ideas

We saw in section 7.3 that in 1922 Einstein pointed out four requirements which should be present in any theory incorporating Mach's principle:

1. The inertia of a body must increase when ponderable masses are accumulated in its neighbourhood.

2. A body must experience an accelerating force when neighbouring masses are accelerated, and, in fact, the force must be in the same direction as that acceleration.

3. A rotating hollow body must generate inside of itself a 'Coriolis field', which deflects moving bodies in the sense of the rotation, and a radial centrifugal field as well.

4. A body in an otherwise empty universe should have no inertia. Related to this is the statement that *all* the inertia of any body should arise from its interaction with other masses in the universe.

As we have seen, these four consequences are not fully implemented in Einstein's general theory of relativity. Here we show that all of them are completely implemented in relational mechanics [67] and [12, Chapter 7].

We begin with the first consequence. Let us suppose a body of gravitational mass m_g interacting with anisotropic distributions of matter and with the isotropic distribution of distant galaxies around it. The force exerted by this anisotropic distribution of matter composed of N bodies is represented by $\vec{F}_{Am} = \sum_{j=1}^{N} \vec{F}_{jm}$, Figure 8.6.

As we have seen, in this case the equation of motion derived in relational mechanics is given by Eq. (8.33), namely

$$\sum_{j=1}^{N} \vec{F}_{jm} - \Phi m_g \vec{a}_{mU} = 0 \, , \tag{8.52}$$

where the acceleration $\vec{a}_{mU}$ is the acceleration of m_g relative to the universal frame of reference. This is analogous to Newton's second law of motion with the test body having an inertial mass m_i given by $m_i = m_g$.

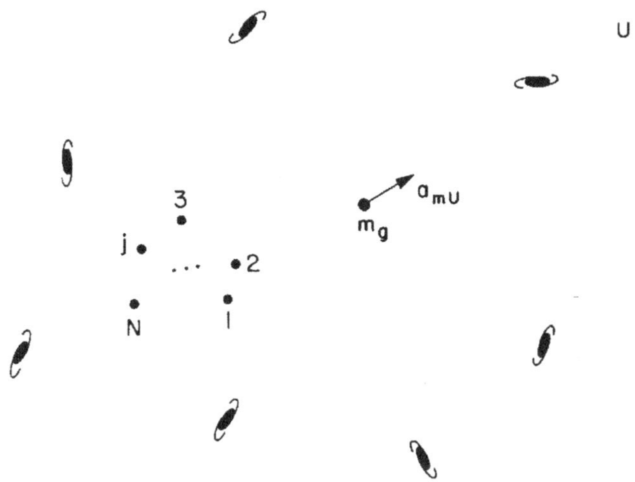

Figure 8.6: Test body interacting with other bodies and galaxies.

We now surround the test body with a spherical shell at rest and without rotation in the universal frame U. The shell has a radius R, thickness dR, and isotropic gravitational matter density ρ_g. The mass of this spherical shell is simply $M_g = 4\pi R^2 dR \rho_g$, shown in Figure 8.7.

We then apply in this second case the third postulate of relational mechanics, the principle of dynamical equilibrium, which states that the sum of all forces acting on m_g is zero. Applying this principle together with Eq. (8.52) and the result (8.12) for the gravitational force exerted by this stationary spherical shell on m_g yields:

$$\sum_{j=1}^{N} \vec{F}_{jm} - H_g \frac{\xi}{3c^2} \frac{m_g M_g}{R} \vec{a}_{mU} - \Phi m_g \vec{a}_{mU} = 0 \, . \tag{8.53}$$

This equation can also be written as (with (8.37)):

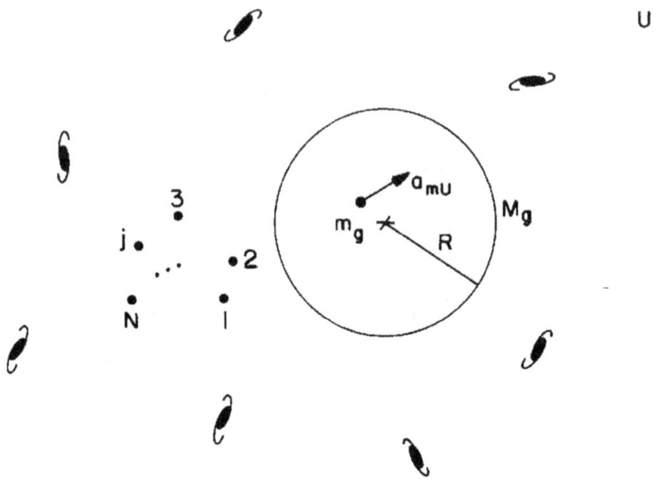

Figure 8.7: The previous situation with the test body surrounded by a spherical shell.

$$\sum_{j=1}^{N} \vec{F}_{jm} = \Phi m_g \left(1 + G \frac{\xi}{3c^2} \frac{M_g}{R} \right) \vec{a}_{mU} \,. \tag{8.54}$$

This is analogous to Newton's second law of motion with the test body having an effective inertial mass given by:

$$m_i = m_g \left(1 + G \frac{\xi}{3c^2} \frac{M_g}{R} \right) \,.$$

This shows that the inertia of a body must increase when ponderable masses are accumulated in its neighbourhood, as required by Mach's principle and as correctly pointed out by Einstein. This is implemented in relational mechanics, but not in Eintein's general theory of relativity.

Let us now analyse the second consequence. To simplify the analysis we consider a one dimensional motion in which two gravitational masses m_{g1} and m_{g2} are interacting through Weber's law, Eq. (8.4). We consider bodies 1 and 2 moving along the x axis, with $x_1 < x_2$, so that $\hat{r}_{12} = - \hat{x}$ (see Figure 8.8).

We simplify the notation utilizing the fact that $r_{12} = |\vec{r}_1 - \vec{r}_2| = |x_1 - x_2| \equiv r$. Moreover, $\vec{r}_{12} = (x_1 - x_2)\hat{x}$, $\vec{v}_{12} = (\dot{x}_1 - \dot{x}_2)\hat{x}$ and $\vec{a}_{12} = (\ddot{x}_1 - \ddot{x}_2)\hat{x}$. This means that $\dot{r}_{12} = \hat{r}_{12} \cdot \vec{v}_{12} = -(\dot{x}_1 - \dot{x}_2)$, and also that $r_{12}\ddot{r}_{12} = \vec{v}_{12} \cdot \vec{v}_{12} - (\hat{r}_{12} \cdot \vec{v}_{12})^2 +$

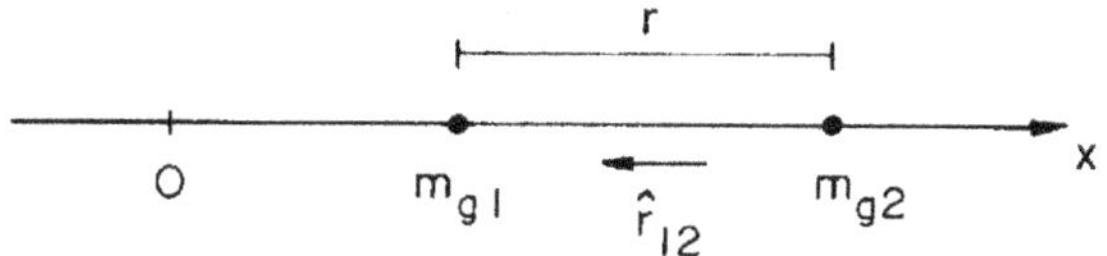

Figure 8.8: Two masses interacting along the x axis.

$\vec{r}_{12} \cdot \vec{a}_{12} = (x_1 - x_2)(\ddot{x}_1 - \ddot{x}_2)$. The force exerted by 2 on 1 is then simplified to the form:

$$\vec{F}_{21} = + \, H_g m_{g1} m_{g2} \frac{\hat{x}}{r^2} \left\{ 1 - \frac{\xi}{c^2} \left[\frac{(\dot{x}_1 - \dot{x}_2)^2}{2} - (x_1 - x_2)(\ddot{x}_1 - \ddot{x}_2) \right] \right\} \; . \quad (8.55)$$

This means that if m_{g2} is accelerated to the right ($\ddot{x}_2 > 0$) there will be a component of the force acting on m_{g1} proportional to $\ddot{x}_2$, namely:

$$\frac{H_g m_{g1} m_{g2} \xi (x_2 - x_1) \ddot{x}_2 \hat{x}}{c^2 r^2} \; .$$

As $\xi > 0$ and $(x_2 - x_1) > 0$, this force points to the right, that is, in the same direction as the acceleration of m_{g2}. If m_{g2} were accelerated to the left, there would also appear a force on body 1 pointing to the left and proportional to $\ddot{x}_2$.

We have shown that in relational mechanics a body experiences an accelerating force when neighbourring masses are accelerated. Moreover, we have shown that this force is in the same direction as the acceleration of the neighbouring masses. All of this is required in order to implement quantitatively Mach's principle, and this was correctly pointed out by Einstein. The derivation of this effect in relational mechanics is extremely simple and natural. A similar result can also be derived in Einstein's general theory of relativity, but in a more complicated way.

Let us now analyse the third consequence. Suppose we are in a frame of reference S in which there is a spherical shell of gravitational mass dM_g with its center at rest relative to S and coinciding with its origin O. Moreover, suppose this spherical shell to be spinning relative to this frame with angular velocity $\vec{\omega}_{MS}$. The gravitational mass of the shell is given by $M_g = 4\pi R^2 dR \rho_g$, where R is its radius, dR its thickness and ρ_g its gravitational mass density. Suppose now there is a test body inside the shell moving relative to this frame. The gravitational force exerted by the shell on the test body is found to be given in relational mechanics by Eq. (8.12), namely:

$$dF = -G \frac{\xi}{3c^2} \frac{m_g M_g}{R} \left[\vec{a}_{mS} \right.$$

$$\left. + \vec{\omega}_{MS} \times (\vec{\omega}_{MS} \times \vec{r}_{mS}) + 2\vec{v}_{mS} \times \vec{\omega}_{MS} + \vec{r}_{mS} \times \frac{d\vec{\omega}_{MS}}{dt} \right] .$$

This shows that in relational mechanics a rotating hollow body generates inside itself a Coriolis force proportional to $2m_g \vec{v}_{mS} \times \vec{\omega}_{MS}$ which deflects moving bodies in the sense of the rotation, and a radial centrifugal force proportional to $m_g \vec{\omega}_{MS} \times (\vec{\omega}_{MS} \times \vec{r}_{mS})$. This is in complete agreement with Mach's principle. As we have seen, the analogous effect in general relativity was derived by Thirring, but with wrong (not observed experimentally) coefficients in front of the Coriolis and centrifugal forces. Moreover, general relativity predicts spurious effects not found in any experiment (the axial term proportional to $\vec{\omega}$).

Let us now analyse the fourth consequence. It also follows immediately from relational mechanics. The inertia of a body, namely, the force $-m_g \vec{a}_{mU}$, was obtained only supposing the contribution from the distant galaxies. If we make these galaxies disappear, which is analogous to making $\rho_o = 0$ in Eqs. (8.12), (8.16) and (8.34), there will be no force analogue to the Newtonian $m_i \vec{a}$. The inertia of the body disappears.

Another way of observing this consequence in relational mechanics is that all forces in this theory are based on two-body interactions. There is no force on any body from "space." It is then meaningless to speak of the motion of a single body in an otherwise empty universe. The simplest system we can consider with motion is a universe composed of two particles.

As we have seen, this does not happen in Einstein's general theory of relativity in which a body in an otherwise empty universe is endowed with its full inertia.

Chapter 9

Applications of Relational Mechanics

In this chapter we present some applications of relational mechanics. When we consider the weight of a body on the earth's surface, it is no longer given simply by $\vec{P} = m_g\vec{g}$, with $\vec{g} = -GM_{gt}\hat{r}/R_t^2$, as this latter expression was obtained with Newton's law of gravitation. Now we have replaced it by Weber's law applied to gravitation. In order to know the force exerted by the earth on a test body moving near its surface we need to integrate Eq. (8.14) for the whole earth. But in this chapter we will only consider situations in which $v^2/c^2 \ll 1$, v being the velocity of the test particle relative to the earth, and $ra/c^2 \ll 1$, r being the distance of the test particle to the center of the earth and a its acceleration relative to the earth. Usually we will be in the frame of the earth, so that $\vec{\omega} = 0$.

In these approximations the force exerted by the earth on a test particle moving near its surface according to Weber's law reduces to a result similar to Newton's, namely: $\vec{P} = m_g\vec{g}$, with $\vec{g} = -H_g M_{gt}\hat{r}/R_t^2$. The only difference is the appearance of H_g instead of the universal constant of gravitation G.

9.1   Uniform Rectilinear Motion

The equation of motion of relational mechanics in a frame of reference in which the set of distant galaxies is at rest and without rotation is given by Eq. (8.33):

$$\sum_{j=1}^{N} \vec{F}_{jm} - \Phi m_g \vec{a}_{mU} = 0 \; . \tag{9.1}$$

If $\sum_{j=1}^{N} \vec{F}_{jm} = 0$ this equation leads to $\vec{a}_{mU} = 0$, observing that $\Phi \neq 0$ and $m_g \neq 0$:

$$\vec{v}_{mU} = \frac{d\vec{r}_{mU}}{dt} = constant \; .$$

The difference as regards Newtonian mechanics is that the velocity $\vec{v}_{mU}$ here is the velocity of the test particle m_g relative to this frame of reference defined by the distant galaxies. Obviously, if the velocity of the test body is constant in this frame it will also be constant in any other frame which moves with a constant velocity relative to the frame of distant galaxies (the universal frame of reference).

If we are disregarding accelerations of the order of 10^{-10} m/s^2 (the typical acceleration of the solar system around the center of our galaxy relative to the frame of distant galaxies) we may say that the test body will move with a constant velocity relative to the frame of fixed stars or to any other frame that moves with a constant linear velocity relative to the set of fixed stars. If we are disregarding accelerations of the order of 10^{-3} to 10^{-2} m/s^2 (the typical centripetal accelerations due to the anual translation and diurnal rotation of the earth relative to the frame of fixed stars) we may say that the test body will move with a constant velocity relative to the earth or to any other frame of reference moving with a constant velocity relative to the earth.

9.2   Constant Force

We now consider the situation in which Eq. (8.48) is satisfied, so that we only consider the acceleration of the test body relative to the earth's surface. In this approximation the acceleration of the test body relative to the set of distant galaxies will be essentially the same as its acceleration relative to the frame of fixed stars and to the earth: $\vec{a}_{mU} \approx \vec{a}_{mf} \approx \vec{a}_{me}$. The equation of motion is then given by (8.49):

$$\sum_{j=1}^{N} \vec{F}_{jm} - \Phi m_g \vec{a}_{me} = 0 \; . \tag{9.2}$$

We now have a situation in which $\sum_{j=1}^{N} \vec{F}_{jm} = \vec{F}_o = constant$. Application of Eq. (9.2) leads to:

$$\vec{a}_{me} = \frac{d\vec{v}_{me}}{dt} = \frac{\vec{F}_o}{\Phi m_g} = constant \; . \tag{9.3}$$

Once more, the difference as regards Newtonian mechanics is that this is not the acceleration of the test body relative to absolute space or relative to an inertial frame of reference, but relative to the earth's surface (in the approximation being considered here, where Eq. (8.48) is valid).

If we are in the frame of reference S' in which the universe as a whole (the set of distant galaxies) has a translational acceleration $\vec{a}_{US'}$ relative to S' and no rotation, the equation of motion of relational mechanics takes the form (8.29):

$$\sum_{j=1}^{N} \vec{F}_{jm} - \Phi m_g(\vec{a}_{mS'} - \vec{a}_{US'}) = 0 \ . \tag{9.4}$$

In the case of a constant anisotropic force, $\sum_{j=1}^{N} \vec{F}_{jm} = \vec{F}_o = constant$, we obtain:

$$\vec{F}_o + \Phi m_g \vec{a}_{US'} - \Phi m_g \vec{a}_{mS'} = 0 \ . \tag{9.5}$$

Here $\vec{a}_{mS'}$ is the acceleration of the test body in this frame. In a frame fixed with the test body, $\vec{a}_{mS'} = 0$, so that the constant force $\vec{F}_o$ is balanced by the gravitational force exerted by the distant accelerated universe, namely, $\vec{F}_o = -\Phi m_g \vec{a}_{US'}$.

9.2.1 Free Fall

Here we study the motion of a test body, such as an apple, falling freely near the surface of the earth (neglecting air friction). When studying the problem of free fall we will neglect the acceleration of the earth (due to the gravitational attraction of the apple) relative to the frame of distant galaxies as compared with the acceleration of the test body in this same frame, due to the negligible mass of the test body (an apple, for instance) compared to the earth's mass.

If the constant force is the weight of the test body due to the gravitational attraction of the earth near its surface, we have: $\vec{F}_o = m_g \vec{g}$, where $\vec{g} = -H_g M_{gt} \hat{r}/R_t^2$ is the gravitational field of the earth. Utilizing this and Eq. (8.16) in Eq. (9.3) yields:

$$\vec{a}_{me} = \frac{m_g \vec{g}}{\Phi m_g} = \frac{3H_o^2}{2\pi \xi \rho_o} \frac{M_{gt} \hat{r}}{R_t^2} = constant \ . \tag{9.6}$$

This explains clearly the observational fact, due to Galileo, that all bodies fall in vacuum with the same acceleration near the surface of the earth, no matter what their weight, form, chemical composition, *etc.* This is due to the fact that the force $-\Phi m_g \vec{a}_{me}$ of relational mechanics is a real force caused by the *gravitational* interaction of the test body with the distant masses in the cosmos. So the mass which appears in $-\Phi m_g \vec{a}_{me}$ is the same mass which appears in the weight of the test body due to its gravitational interaction with the earth, namely: $\vec{P} = m_g \vec{g}$. The explanation of this remarkable fact which so puzzled Galileo, Newton, Einstein and many others, becomes obvious. This explanation of relational mechanics is very simple and elegant. We don't need to postulate this proportionality, as had been done in general relativity. Instead of postulating it (without a better understanding of the fact), this result is *derived* in relational mechanics. We then acquire a complete understanding of this fact, which opens up many new possibilities.

The graphical representation of the forces in this case is given in Figure 9.1. The accelerations in this frame are represented in Figure 9.2.

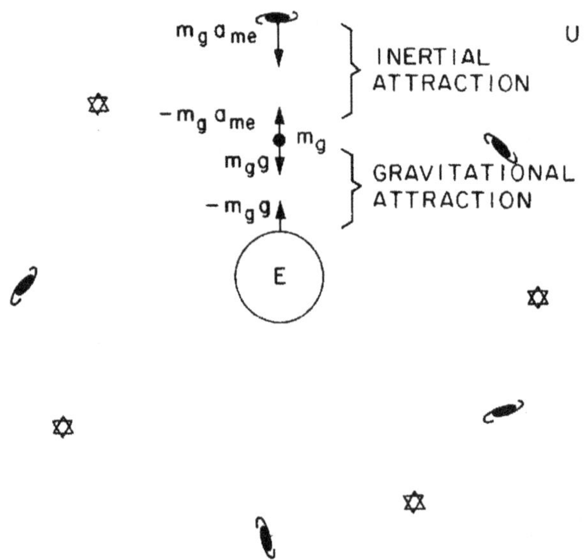

Figure 9.1: Forces in the situation of free fall in the frame of the earth.

There is action and reaction in the gravitational interaction between the earth and the test body, as there is action and reaction in the gravitational interaction between the distant universe and the test body (here called inertial attraction to highlight the identification and new meaning of Newtonian mechanics). The weight of the test body is paired with an opposite force acting on the earth, while the gravitational force exerted by the distant universe on m_g, $-\Phi m_g \vec{a}_{me}$, is paired with an opposite force acting on the distant universe (distributed between all galaxies). In Newtonian mechanics the inertial forces, such as $-m_i \vec{a}$, were not associated with a counter-force on another body. The

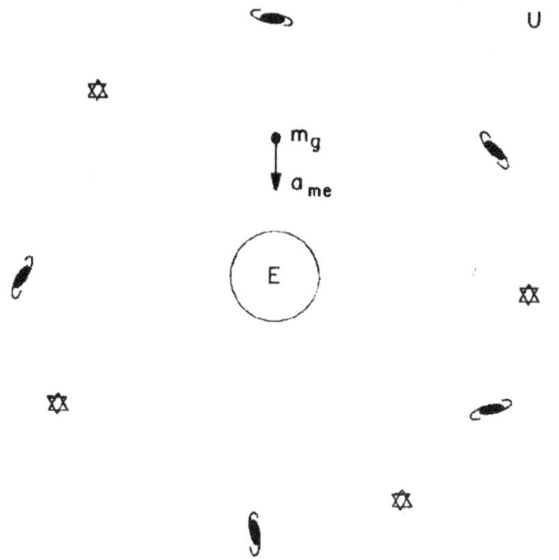

Figure 9.2: Accelerations as seen from the earth.

forces acting on the test body, its weight $(m_g\vec{g})$ and the force exerted by the stationary distant universe on the accelerated test body $(-\Phi m_g\vec{a}_{me})$ balance one another yielding the acceleration of the body.

The fact that the inertial forces ($m_i\vec{a}$, centrifugal, Coriolis, *etc.*) of classical mechanics do not comply with the law of action and reaction was clearly pointed out by Einstein in the Foreword (1953) of the book *Concepts of Space*, by Max Jammer: "The concept of space was enriched and complicated by Galileo and Newton, in that space must be introduced as the independent cause of the inertial behavior of bodies if one wishes to give the classical principle of inertia (and therewith the classical law of motion) an exact meaning. To have realized this fully and clearly is in my opinion one of Newton's greatest achievements. In contrast with Leibniz and Huygens, it was clear to Newton that the space concept (a) [space as positional quality of the world of material objects; as opposed to concept (b): space as container of all material objects] was not sufficient to serve as the foundation for the inertia principle and the law of motion. He came to this decision even though he actively shared the uneasiness which was the cause of the opposition of the other two: space is not only introduced as an independent thing apart from material objects, but also is assigned an absolute role in the whole causal structure of the theory. This role is absolute in the sense that space (as an inertial system) acts on all material objects, while these do not in turn exert any reaction on space," [33, pp. xiii-xiv]. This can be seen clearly in the bucket experiment. If the bucket and water are at rest or move uniformly in a straight line relative to the Newtonian absolute space, the water surface is flat. When the water is spinning relative to absolute space its surface is concave, rising towards the sides of the bucket. We can say that absolute space is acting on the water, pressing it towards the sides of the bucket. On the other hand, nothing happens to absolute space. The spinning water does not exert any reaction force on space. In relational mechanics this no longer happens. All forces termed inertial in Newtonian mechanics are now considered to be caused by an acceleration between the test body and the distant universe. And there is a reaction to these forces, exerted by the test body on the distant universe (distributed between all galaxies).

If we are in the frame of reference S' of the test body so that it does not move in this frame (only the earth approaches the test body), we have, from Eq. (9.5) with $\vec{a}_{mS'} = 0$ and remembering that $\vec{F}_o = m_g\vec{g}$:

$$m_g\vec{g} + \Phi m_g\vec{a}_{US'} = 0 \ .$$

The forces in this case are represented in Figure 9.3, while the accelerations in this frame are seen in Figure 9.4.

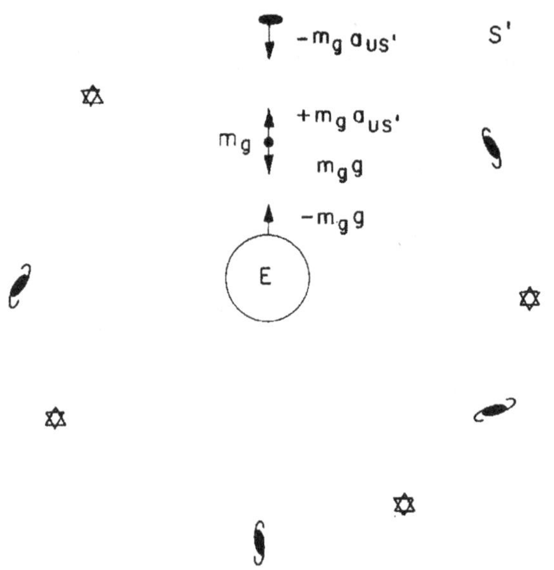

Figure 9.3: Forces in the situation of free fall in the frame of the test body.

What we see here is again action and reaction between the earth and the test body, and the same between the distant universe and the test body. But

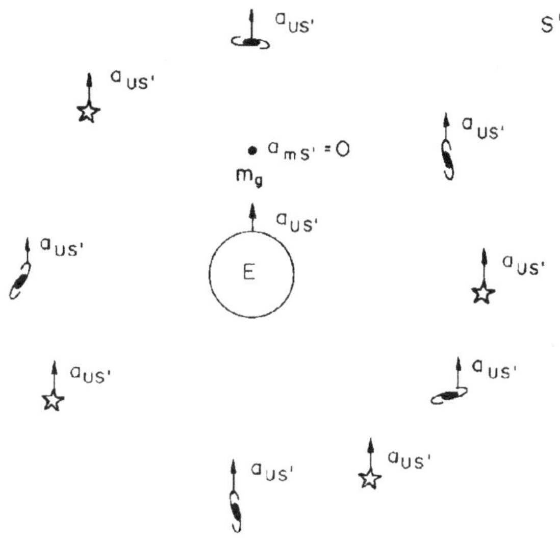

Figure 9.4: Accelerations as seen by the test body.

now the interpretation is a little bit different. Now we say that the weight of the body is balanced by an upward gravitational force exerted by the accelerated universe, so that the test body does not move in its own frame of reference.

We now show another fundamental aspect of relational mechanics which is not implemented in Newtonian mechanics or in Einstein's theories of relativity, namely, the relative aspect of mass or density. In classical mechanics or in general relativity the acceleration of free fall near the surface of the earth is given by

$$a^N = a^E = G\frac{M_t}{R_t^2} \, , \tag{9.7}$$

where $G = 6.67 \times 10^{-11}$ Nm^2/kg^2 is the universal constant of gravitation. Utilizing the earth's mass and radius we obtain in this first case for Newton and Einstein: $a_I^N = a_I^E = 9.8$ m/s^2. If we double the masses of all bodies in the universe (the test body, the earth, the stars and distant galaxies), keeping all distances and sizes fixed (which is equivalent to saying that we are doubling the densities of all bodies), this equation predicts that the acceleration of free fall will double: $a_{II}^N = a_{II}^E = 19.6$ m/s^2. This shows that not only space and time are absolute in these theories (without relation to anything external), but the same also applies to mass. But this is against Mach's ideas. Remember that he wrote: "All masses and all velocities, and consequently all forces, are relative" [39, p. 279]. This means that if we doubled or halved all masses, it should be impossible to discover this. But here we showed that in the theories of Newton and Einstein, if we double all the masses the acceleration of free fall also doubles, so that by this effect we might discover that the absolute masses of the universe had doubled.

On the other hand, in relational mechanics the acceleration of free fall is given by Eq. (9.6). Writing the mass of the earth as $M_{gt} = 4\pi R_t^3 \rho_t/3$, where ρ_t is the mean gravitational density of the earth's mass, we have:

$$a^{RM} = \frac{2}{\xi} R_t H_o^2 \frac{\rho_t}{\rho_o} \ . \tag{9.8}$$

Utilizing the known values of R_t, H_o, ρ_t and ρ_o yields $a^{RM} \approx 9.8$ m/s^2. This equation shows that the acceleration of free fall according to relational mechanics depends only on the ratio of the matter density of the earth by the average matter density of the universe. If we double or halve both of them, the acceleration of free fall should remain the same. The explanation is very simple: When we double the mass of the earth, the weight of the test body doubles, but when we also double the mass of the distant universe, it also doubles the "inertia" of the test body (utilizing the language of Newtonian mechanics). In other words, the force $-\Phi m_g \vec{a}_{mU}$ exerted by the distant universe on m_g also doubles. For this reason the net acceleration of free fall of the test body will remain the same, 9.8 m/s^2.

We now make a variation of this example. In the first case we have a test body of 1 kg falling freely near the earth's surface. All three theories (Newtonian, Einsteinian and relational mechanics) predict the observed value: $a_I^N = a_I^E = a_I^{RM} = 9.8$ m/s^2. In the second case we double the mass of the test body, keeping the masses of the earth, stars and galaxies constant. Once more all three theories predict, by Eqs. (9.7) and (9.8): $a_{II}^N = a_{II}^E = a_{II}^{RM} = 9.8$ m/s^2. This is confirmed by Galileo's experiment. In the third situation we keep the original mass of 1 kg for the test body but halve the masses of the Earth, stars and galaxies, keeping all dimensions and distances constant. According to Eq. (9.7) we conclude that classical mechanics and general relativity predict an acceleration of free fall given by: $a_{III}^N = a_{III}^E = 4.9$ m/s^2. Only relational mechanics, by Eq. (9.8), predicts the same acceleration in this case: $a_{III}^{RM} = 9.8$ m/s^2. This is the only prediction which makes sense philosophically. The reason is that the ratio of any two masses in case II is the same as the ratio of the equivalent masses in case III, so that it should be impossible to distinguish these two cases. We will represent the test body by the superscript a, the earth by e and the galaxies by g. As $m_{II}^a/m_{II}^e = m_{III}^a/m_{III}^e$, $m_{II}^a/m_{II}^g = m_{III}^a/m_{III}^g$ and $m_{II}^e/m_{II}^g = m_{III}^e/m_{III}^g$ we should have the acceleration of free fall in case II equal to the acceleration of free fall in case III. This is predicted only by relational mechanics.

This shows that in classical mechanics and in Einstein's general theory of relativity there is not only absolute space and time but also absolute mass. The reason for this is Newton's theorem that a spherical shell exerts no force on internal test particles, no matter what the position or motion of the test particle. The same happens in general relativity, as was shown by Brans in 1962 [116], [75] and [117]. Although the acceleration of free fall depends on the mass of the earth in these theories, it does not depend on the mass of the stars and galaxies. This means that we can remove spherical shells composed of stars and galaxies around the earth without changing the acceleration of a falling apple. This does not happen in relational mechanics, as in this case the acceleration of free fall will depend on the ratio of the mass (density) of the earth to the mass (density) of the distant universe.

9.2.2 Charge Moving Inside an Ideal Capacitor

The other example analysed here is that of a charged particle q which suffers a force from an ideal capacitor which is at rest relative to the earth. The capacitor is charged with surface densities $\pm\sigma$ on its plates orthogonal to the z axis located at $\pm z_o$, as in Figure 2.3. The test charge is located at time t at $\vec{r} = z\hat{z}$ and moves with velocity $\vec{v} = v_x\hat{x} + v_y\hat{y} + v_z\hat{z}$ and acceleration $\vec{a} = a_x\hat{x} + a_y\hat{y} + a_z\hat{z}$ relative to the plates of the capacitor.

The force on a test charge inside the ideal capacitor in Weber's electrodynamics is different from Lorentz's force in this case. Weber's force is given by [12, Section 6.7]:

$$\vec{F}(-z_o < z < z_o) = -4\pi H_e q\sigma \left[\hat{z} + \frac{v^2}{2c^2}\hat{z} \right.$$

$$\left. - \frac{v_x(v_x\hat{x} + v_y\hat{y})}{c^2} - \frac{z\vec{a}}{c^2} + \frac{2za_z}{c^2}\hat{z} \right] \, ,$$

Here we consider only the approximation in which $v^2/c^2 \ll 1$ and $za/c^2 \ll 1$, so that this result reduces to the classical one based on Lorentz's force, namely: $\vec{F}_o \approx -4\pi H_e q\sigma\hat{z} = q\vec{E}$.

Eqs. (9.2) and (8.43) lead to:

$$\vec{a}_{me} = \frac{q}{\Phi m_g}\vec{E} = -\frac{6}{\xi}\frac{H_e}{H_g}\frac{H_o^2}{\rho_o}\frac{q\sigma}{m_g}\hat{z} = -\frac{q\sigma}{m_g\varepsilon_o}\hat{z} \, .$$

As there is no relation between the electrical charge q and the gravitational mass of the particle m_g, the acceleration of two different particles, such as an alpha particle and a proton in the same capacitor, may be different, as is indeed the case.

If we doubled (halved) the density of external galaxies, maintaining the test charge and capacitor without alteration, the acceleration of the test charge would halve (double), as we can see from the dependence of a_{me} on ρ_o. This is predicted by relational mechanics but not by Newtonian nor Einsteinian theories. Once more this is explained in relational mechanics by saying that the effective inertial mass (employing the common terms of classical mechanics) will double (halve) in this case.

9.2.3 Accelerated Train

We now consider a train with a constant acceleration relative to the earth's surface, with a simple pendulum fixed on its roof. We analyse here only the equilibrium situation when the bob of the pendulum does not move relative to the train, so that it is at a constant angle θ relative to the vertical. The forces acting on the bob, neglecting air resistance, are its weight $\vec{P} = m_g \vec{g}$, the tension in the string, $\vec{T}$, and the gravitational force due to the distant universe. Eq. (9.2) leads to:

$$m_g \vec{g} + \vec{T} - \Phi m_g \vec{a}_{me} = 0 \ , \tag{9.9}$$

where in relational mechanics $g = H_g M_{gt}/R_t^2$.

Utilizing the angle θ of Figure 2.5 yields:

$$m_g g = T \cos \theta \ ,$$

$$T \sin \theta = \Phi m_g a_{me} \ .$$

From these expressions we obtain immediately, with Eq. (8.16) and $M_{gt} = 4\pi R_t^3 \rho_t / 3$:

$$\tan \theta = \frac{\Phi a_{me}}{g} = \frac{\xi}{2} \frac{\rho_o}{\rho_t} \frac{a_{me}}{R_t H_o^2} \ . \tag{9.10}$$

Hence it follows immediately from relational mechanics that the angle of inclination will depend only on the ratio of the linear acceleration of the body relative to the earth, a_{me}, and the free fall acceleration, g. It will not depend on the weight, form or chemical composition of the body, although the tension in the string will depend on the weight of the test body.

We now analyse this problem in the train's frame of reference T. In this case, the train and the pendulum are seen at rest, while the earth and the distant universe are seen moving with acceleration $\vec{a}_{UT}$ given by: $\vec{a}_{UT} = -\vec{a}_{te}$, where $\vec{a}_{te}$ is the acceleration of the train relative to the earth's surface. Eq. (9.4) now takes the form:

$$\vec{P} + \vec{T} + \Phi m_g \vec{a}_{UT} = \Phi m_g \vec{a}_{mT} .$$

Since in this frame the test body is at rest, $\vec{a}_{mT} = 0$, and $\vec{a}_{UT} = -\vec{a}_{te}$, we obtain the same equation as Eq. (9.9), so that the final result is the same. Only the interpretation is now slightly different from the situation seen in the earth's frame. Now we say that the tension $\vec{T}$ and the weight $\vec{P}$ are balanced by a gravitational force exerted by the accelerated distant universe on m_g, so that m_g does not move in this frame. See Figure 9.5.

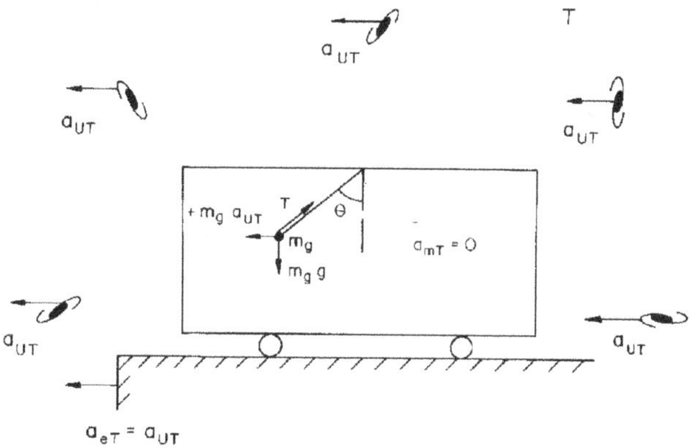

Figure 9.5: Accelerations and forces as seen from the train (T). The earth and the galaxies have an acceleration a_{UT} to the left while the body of mass m_g has no acceleration in this frame. The accelerated galaxies exert a force $m_g a_{UT}$ to the left on m_g, which balances the weight of the body and the tension in the string.

The easiest way to understand the equilibrium situation in the train's frame of reference is to think that the universe accelerated to the left exerts a gravitational force on the bob pointing to the left proportional to m_g and to the acceleration of the universe. Given the universe acceleration, we can then find the angle θ and the tension T in the string such that the body will be at rest. In other words, we equate $T \cos \theta$ with $P = m_g g$ and $T \sin \theta$ with $\Phi m_g a_{UT}$. In this way we find θ and T such that the bob is at rest, $a_{mT} = 0$.

From Eq. (9.10) we can see that the angle of inclination is proportional to ρ_o/ρ_t, *i.e.* to the ratio of the average density in the universe by the density of the earth. Keeping R_t constant and always analysing the situation in which the train has the same acceleration relative to the earth, $a_{me} = constant$, we conclude that $\tan\theta$ will double if ρ_o/ρ_t doubles. This will happen no matter if we halve ρ_t, keeping ρ_o constant, or if we double ρ_o, keeping ρ_t constant. Once more, only relational mechanics arrives at this simple and appealing result.

9.3   Harmonic Oscillator

We consider here bodies moving over the surface of the earth such that condition (8.48) is satisfied. In this case, the equation of motion of relational mechanics reduces to Eq. (9.2).

9.3.1   Spring

We first analyse a body fastened to a spring and oscillating horizontally. Its weight is balanced by the reaction force exerted by a frictionless table, shown in Figure 2.6. In the one-dimensional motion of a test body connected to a spring of elastic constant K the equation of motion reduces to:

$$- Kx - \Phi m_g a_{me} = 0 \ .$$

Here x is the displacement of the body from the equilibrium position. As we are in a one-dimensional problem we can put $a_{me} = \ddot{x}$. The solution of this equation is then found to be:

$$x(t) = A_o \sin(\omega t + \theta_o) \ ,$$

where A_o is the amplitude of oscillation (specified by the initial conditions), θ_o is the phase of oscillation (also specified by initial conditions) and ω is the frequency of oscillation given by (with Eq. (8.16)):

$$\omega = \sqrt{\frac{K}{\Phi m_g}} = \sqrt{\frac{3}{2\pi\xi} \frac{K}{m_g} \frac{H_o^2}{H_g \rho_o}} \ .$$

This result shows that for springs oscillating horizontally the frequency of vibration is inversely proportional to the weight of the test body, as observed experimentally. Comparison with Eq. (2.11) of Newtonian mechanics shows that the greatest difference is the appearance in relational mechanics of the gravitational mass m_g instead of the inertial mass m_i.

Doubling the density of galaxies in the universe, while keeping the spring and test body unaltered, would decrease the frequency of oscillation by $\sqrt{2}$. This would be the same as doubling the Newtonian inertial mass of the test body.

9.3.2 Simple Pendulum

We now have a simple pendulum oscillating near the earth's surface. Neglecting air resistance, the forces acting on the bob of gravitational mass m_g are its weight $\vec{P} = m_g\vec{g} = m_g(H_g M_{gt}/R_t^2)\hat{g}$, the tension in the string $\vec{T}$ and the gravitational force exerted by distant galaxies, $-\Phi m_g \vec{a}_{me}$:

$$m_g\vec{g} + \vec{T} - \Phi m_g \vec{a}_{me} = 0 \ .$$

The length ℓ of the string is a constant. Using a polar coordinate system with $\vec{T} = -T\hat{\ell}$, $\vec{a}_{me} = -(\ell\dot{\theta}^2)\hat{\ell} + \ell\ddot{\theta}\hat{\theta}$, Figure 2.7, yields:

$$T - m_g g\cos\theta - \Phi m_g \ell\dot{\theta}^2 = 0 \ ,$$

$$-m_g g\sin\theta - \Phi m_g(\ell\ddot{\theta}) = 0 \ .$$

This last equation shows that even without further approximations the value of the angle of oscillation as a function of time will not depend on m_g, as it cancels out of the expression. The same did not happen in the previous equation for the tension T.

In the approximation of small oscillations ($\theta \ll \pi/2$) this last equation and its solution reduce to (with Eq. (8.16) and $M_{gt} = 4\pi R_t^3 \rho_t/3$):

$$\ddot{\theta} + \frac{g}{\Phi\ell}\theta = \ddot{\theta} + \frac{2}{\xi}H_o^2\frac{R_t}{\ell}\frac{\rho_t}{\rho_o} = 0 \ ,$$

$$\theta = A_1\cos(\omega t + B_1) \ ,$$

where A_1 and B_1 are constants depending on the initial conditions and ω is the frequency of oscillation given by:

$$\omega = \sqrt{\frac{g}{\Phi\ell}} = \sqrt{\frac{2}{\xi}H_o^2\frac{R_t}{\ell}\frac{\rho_t}{\rho_o}} \ .$$

Relational mechanics explains at once Newton's experimental result that bodies of different chemical composition oscillate with the same frequency in pendulums of the same length in the same location at the earth's surface.

Doubling the number of galaxies in the universe (*i.e.*, doubling ρ_o), while keeping the string, earth and the test body unaltered, would decrease the frequency of oscillation by $\sqrt{2}$. It would be the same as halving ρ_t, while keeping R_t and ρ_o constants. It would also be the same as doubling the Newtonian inertial mass of the test body, while keeping its gravitational mass unaltered.

9.4   Uniform Circular Motion

9.4.1   Circular Orbit of a Planet

In this subsection we consider two bodies orbiting around one another relative to distant masses due to their gravitational attraction. The frame of reference considered here is that of the fixed stars, and the bodies might be the sun and a planet. The centripetal acceleration of the solar system around the center of our galaxy is given approximately by $a_s \approx 10^{-10}$ m/s^2. The typical centripetal accelerations of the planets around the sun in the frame of fixed stars is the earth's, namely: 10^{-2} to 10^{-3} m/s^2. As these accelerations are much greater than 10^{-10} m/s^2, in planetary motion we can disregard the acceleration of the solar system relative to distant galaxies. The set of distant galaxies may be considered essentially without acceleration relative to the frame of fixed stars. This means that $\vec{a}_{mU} \approx \vec{a}_{mf}$. So Eq. (8.33) reduces to:

$$\sum_{j=1}^{N} \vec{F}_{jm} - \Phi m_g \vec{a}_{mf} = 0 \ ,$$

where $\vec{a}_{mf}$ is the acceleration of the test body m_g relative to the frame of fixed stars. This is the equation of motion of relational mechanics which is valid in the frame of fixed stars in this approximation where $a_{mf} \gg 10^{-10}$ m/s^2.

The gravitational force exerted between two bodies 1 and 2 is given by Weber's law, Eq. (8.4):

$$\vec{F}_{21} = -H_g m_{g1} m_{g2} \frac{\hat{r}_{12}}{r_{12}^2} \left[1 - \frac{\xi}{c^2} \left(\frac{\dot{r}_{12}^2}{2} - r_{12}\ddot{r}_{12} \right) \right] = -\vec{F}_{12} \ .$$

We consider here the situation in which they are moving in circles relative to the fixed stars, keeping a constant distance to the center of mass. In this case $\dot{r}_{12} = 0$ and $\ddot{r}_{12} = 0$, so that Weber's force reduces to the Newtonian force.

We choose the origin of the coordinate system at the center of mass. The situation seen in the frame of fixed stars is shown in Figure 9.6.

In this case the accelerations of bodies 1 and 2 are only their centripetal accelerations, namely: $\vec{a}_{1f} = -(v_{t1}^2/r_1)\hat{r}_1$ and $\vec{a}_{2f} = -(v_{t2}^2/r_2)\hat{r}_2$. Here v_{t1} is the tangential velocity of body 1 relative to the frame of fixed stars at a distance r_1 to the center of mass and $\hat{r}_1$ is the radial unit vector pointing to it, and analogously for body 2. As $\hat{r}_1 = -\hat{r}_2 = \hat{r}_{12}$ and $\vec{F}_{21} = -\vec{F}_{12}$ we are led to:

$$F = H_g \frac{m_{g1} m_{g2}}{r_{12}^2} = \Phi m_{g1} a_{1f} = \Phi m_{g1} \frac{v_{t1}^2}{r_1} = \Phi m_{g2} a_{2f} = \Phi m_{g2} \frac{v_{t2}^2}{r_2} \ .$$

The gravitational masses cancel as usual. As we have $m_{g1} r_1 = m_{g2} r_2$ we see that both bodies orbit at a constant angular velocity ω_{mf} relative to the fixed stars given by, with Eq. (8.16):

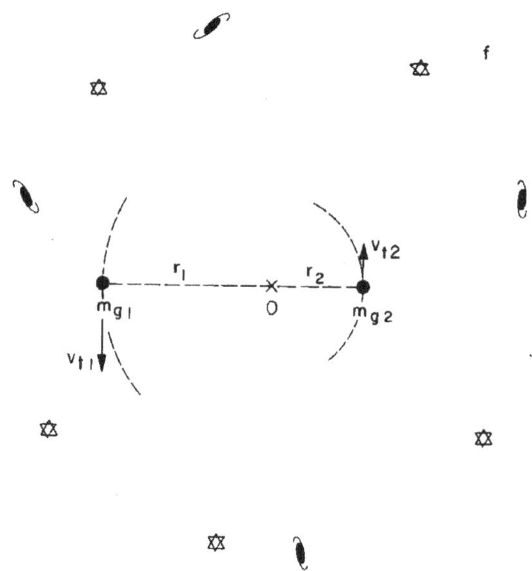

Figure 9.6: Gravitational orbits in the frame of fixed stars.

$$\omega_{mf} = \frac{v_{t1}}{r_1} = \frac{v_{t2}}{r_2} = \sqrt{\frac{3H_o^2}{2\pi\xi\rho_o}\frac{m_{g1}}{r_{12}^2 r_2}} = \sqrt{\frac{3H_o^2}{2\pi\xi\rho_o}\frac{m_{g2}}{r_{12}^2 r_1}} \ .$$

If the z axis is normal to the plane of motion and pointing according to the right hand rule following the planets' motion, then it will point upwards in Figure 9.6. The vectorial angular rotation of the planets will be then given by $\vec{\omega}_{mf} = \omega_{mf}\hat{z}$.

Here we must stress a great conceptual difference when this problem is treated in Newtonian and in relational mechanics. In the first case, it is a simple two body problem, with the planets or bodies orbiting in space. On the other hand, in relational mechanics this is a many body interaction, namely: the two bodies (the sun and a planet, for instance) plus the distant masses in the cosmos, such as the distant galaxies. The distant galaxies play a fundamental role in relational mechanics and cannot be neglected. With the real universe full of bodies we cannot treat any single "two body" problem, as the stars and distant galaxies will exert a real force on any accelerated body. This conceptual difference should always be kept in mind.

Now suppose we are in a frame of reference S centered on the center of mass but in which the planets are at rest. In this frame the distant galaxies and the fixed stars are seen spinning as a whole with an angular velocity (neglecting 10^{-10} m/s^2 compared to ω_{US}):

$$\vec{\omega}_{US} = -\vec{\omega}_{mf} = -\sqrt{\frac{3H_o^2}{2\pi\xi\rho_o}\frac{m_{g1}}{r_{12}^2 r_2}}\,\hat{z}\ .$$

The relational equation of motion in this case is given by Eq. (8.47), namely:

$$\sum_{j=1}^{N} \vec{F}_{jm} - \Phi m_g \left[\vec{a}_{mS} + \vec{\omega}_{US} \times (\vec{\omega}_{US} \times \vec{r}_{mS}) \right.$$

$$\left. + 2\vec{v}_{mS} \times \vec{\omega}_{US} + \vec{r}_{mS} \times \frac{d\vec{\omega}_{US}}{dt} \right] = 0\ .$$

As in this frame the bodies 1 and 2 are seen at rest, $\vec{v}_{mS} = 0$ and $\vec{a}_{mS} = 0$. Moreover, $d\vec{\omega}_{US}/dt = 0$. So this equation reduces to:

$$H_g \frac{m_{g1} m_{g2}}{r_{12}^2} = \Phi m_{g1} \omega_{US}^2 r_{1S}\ ,$$

and an analogous equation for body 2. In other words, the gravitational force between the two bodies 1 and 2 is balanced by a real centrifugal gravitational force exerted by the spinning set of distant galaxies on each one of them. This explains how they can keep a constant distance relative to one another and remain at rest in this frame, despite their gravitational attraction. In Newtonian mechanics this could be explained only by the introduction of a "fictitious" centrifugal force without any known physical origin. In relational mechanics we identify the bodies that are causing this centrifugal force, namely, the distant galaxies. We also identify the origin of this force, namely, a gravitational attraction depending on relative motion, shown in Figure 9.7.

If we could keep the solar system without modifications but could double the number or density of galaxies in the universe, then relational mechanics would predict that the bodies would behave as having twice their present Newtonian inertial masses. This is evident from Eqs. (8.16) to (8.27).

9.4.2   Two Globes

The situation of the two globes connected by a string and spinning relative to the distant galaxies and fixed stars is the same situation as the previous problem, replacing the gravitational force by the tension $\vec{T}$ in the string.

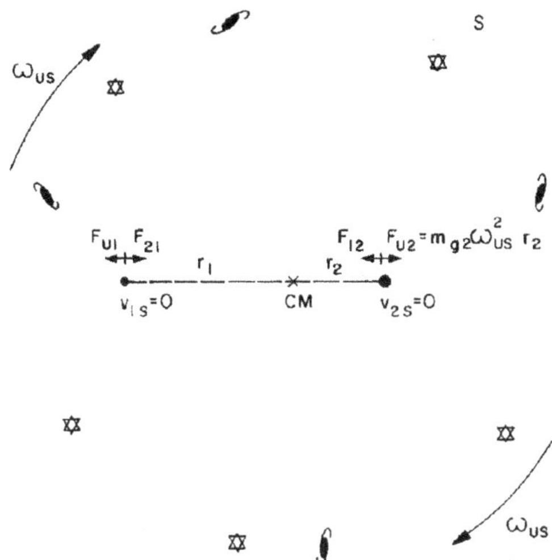

Figure 9.7: Real gravitational centrifugal force balancing the gravitational attraction between two bodies.

In the frame of the fixed stars the globes rotate around the center of mass and the tension in the string is balanced by the gravitational force $-\Phi m_g \vec{a}_{mf}$ due to their acceleration relative to the distant galaxies and fixed stars. See Figure 9.8.

In the frame S in which they are seen at rest the tension in the string is balanced by the centrifugal gravitational force $-\Phi m_g \vec{\omega}_{US} \times (\vec{\omega}_{US} \times \vec{r}_{mS})$ due to the rotation of the distant universe around the globes. See Figure 9.9.

The main difference between relational and Newtonian mechanics in this case is that the tension in the string will always appear provided the relative rotation is the same. In the frame of the fixed stars, f, the set of stars and distant galaxies is essentially at rest while the globes rotate with an angular velocity $\vec{\omega}_{gf} = \omega_o \hat{z}$. In another frame S in which the globes are at rest, it is the set of fixed stars and distant galaxies that rotate relative to S with an angular velocity $\vec{\omega}_{US} = -\omega_o \hat{z}$. As we had seen in subsection 2.4.2, for Newton there would appear no tension in the cord in this last situation. See Figure 9.10.

But for Mach, the tension would be there as in the previous situation. Relational mechanics has implemented Mach's ideas quantitatively showing with Weber's law for gravitation that in this last situation there will appear a real centrifugal force creating or balancing the tension in the string, as in Figure 9.9. Provided the kinematical rotation is the same (globes rotating with $\omega_o \hat{z}$ with galaxies and stars at rest, or the globes at rest with the galaxies and fixed stars rotating with $-\omega_o \hat{z}$), the dynamics will also be exactly the same (the same

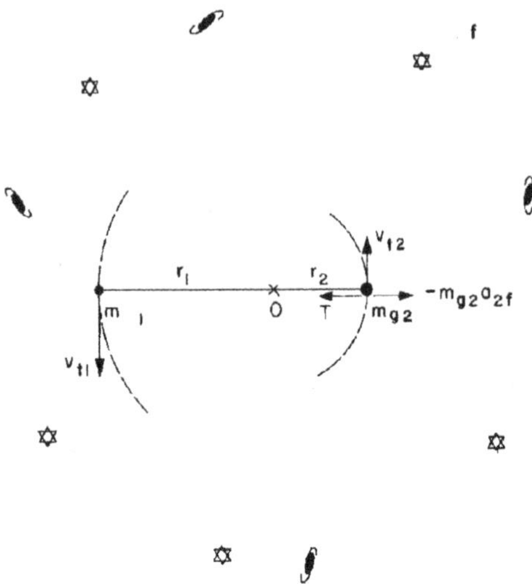

Figure 9.8: Globes rotating relative to the distant galaxies.

tension in the string in both cases). This prediction does not happen in classical mechanics or in Einstein's general theory of relativity. In relational mechanics we cannot know who is really rotating. But from the tension in the string we can conclude that there is a relative rotation between the globes and the distant universe.

Suppose now we double the density of external galaxies, without changing the cord and the globes. It would be more difficult to rotate the globes, due to their increased inertia. But if we rotated the globes with the same angular velocity relative to the distant galaxies, the tension in the cord would double compared with the present situation. We can see this by observing that the tension will be proportional to $\Phi m_g a_{mf}$. As Φ is proportional to ρ_o, doubling the number of galaxies will double the tension.

9.4.3 Newton's Bucket Experiment

We are now in a situation in which condition (8.48) is satisfied, so that the equation of motion of relational mechanics takes the simple form of Eq. (9.2):

$$\sum_{j=1}^{N} \vec{F}_{jm} - \Phi m_g \vec{a}_{me} = 0 .$$

This is similar to Newton's second law of motion, with m_i replaced by m_g and the acceleration relative to an inertial frame replaced by the acceleration

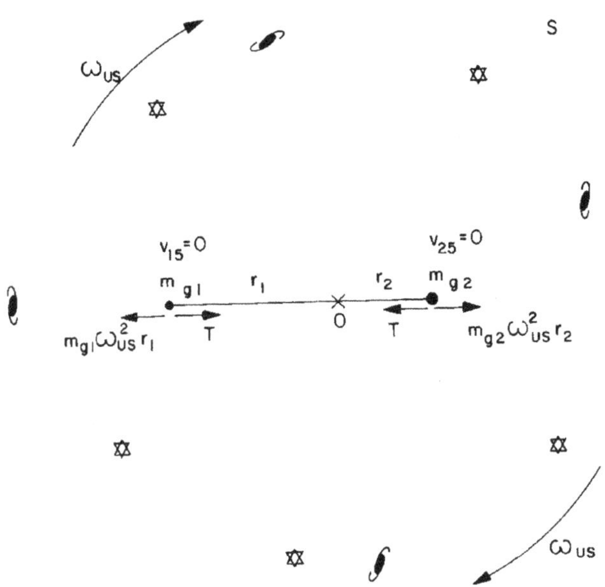

Figure 9.9: Rotating frame of galaxies balancing or creating the tension between the stationary globes.

relative to the earth. Then the solution is the same, namely, the concave surface in the form of a paraboloid of revolution. The procedure to arrive at the equation describing the form of the surface is the same as in subsection 2.4.3:

Let us consider a small volume of liquid $dm_g = \rho_g dV$ just below the surface. It is acted upon by the downward force of gravity, $dP = dm_g g = dm_g(H_g M_{gt}/R_t^2)$, and by a force normal to the surface of the liquid due to the gradient of pressure, dE. This portion of liquid moves in a circle centered on the z axis, so that there is no net vertical force acting on it. The distant galaxies exert only a force pointing towards the z axis changing its direction of motion: $-\Phi dm_g \vec{a}_{we}$, where $\vec{a}_{we}$ is the acceleration at each point of the water relative to the earth. From Figure 2.18 we obtain in this case ($a_c = \omega_{we}^2 x$ being the acceleration of dm_g at a distance x from the axis of rotation, with an angular velocity relative to the earth given by ω_{we}):

$$dE \cos\alpha = dP = dm_g g = dm_g \frac{H_g M_{gt}}{R_t^2} \ , \tag{9.11}$$

$$dE \sin\alpha = \Phi dm_g a_c = \Phi dm_g \omega_{we}^2 x \ . \tag{9.12}$$

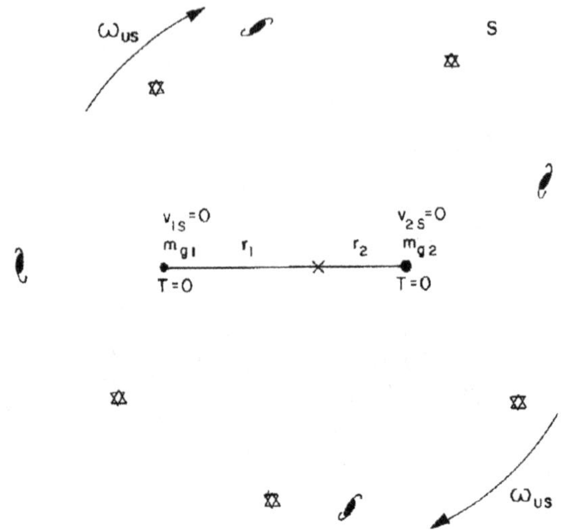

Figure 9.10: According to Newtonian mechanics, a rotating set of stars and galaxies does not produce any centrifugal force.

From these two equations we get:

$$\tan\alpha = \frac{\Phi\omega_{we}^2 R_t^2}{H_g M_{gt}}x \; . \tag{9.13}$$

Utilizing the fact that $\tan\alpha = dz/dx$, where dz/dx is the inclination of the curve at each point, and the fact that we want the equation of the curve which contains the origin $x = z = 0$ yields (taking into account Eq. (8.16) and the fact that $M_{gt} = 4\pi R_t^3 \rho_t/3$):

$$z = \frac{\Phi\omega_{we}^2 R_t^2}{2H_g M_{gt}}x^2 = \left(\frac{\xi}{4}\frac{\omega_{we}^2}{R_t H_o^2}\frac{\rho_o}{\rho_t}\right)x^2 \; . \tag{9.14}$$

In the frame of the bucket, the water is at rest, while the earth, the fixed stars and the distant galaxies are all spinning in the opposite direction (compared with the spinning water seen from the earth's frame of reference). In the frame of the bucket, B, the equation of motion of relational mechanics takes the form of Eq. (8.47). As in this frame the water is at rest, $\vec{v}_{mB} = 0$ and $\vec{a}_{mB} = 0$. Moreover, the angular rotation of the universe is essentially a constant, $d\vec{\omega}_{UB}/dt = 0$. So, this equation reduces to:

$$\sum_{j=1}^{N}\vec{F}_{jm} - \Phi m_g\vec{\omega}_{UB}\times(\vec{\omega}_{UB}\times\vec{r}_{mB}) = 0 \; .$$

This is similar to Newton's second law of motion with centrifugal force when the test body does not move. This means that in this frame relational mechanics predicts the appearance of a real gravitational centrifugal force exerted by the distant universe spinning around the bucket. We can then say that this centrifugal force presses the water against the wall of the bucket making the water rise on this wall until the centrifugal force is balanced by the gradient of pressure.

Let us now suppose we multiply the density of matter ρ_o in the universe by a constant k, leaving the earth, bucket and water without alterations in their masses, sizes or densities. According to Eq. (9.14) we can see that the curvature of the paraboloid (the coefficient in front of x^2) will also be proportional to k. This shows that keeping the same angular rotation of the water relative to the earth and distant galaxies would make it rise more on the sides of the vessel if $k > 1$. If $k \to 0$, which means annihilating the external galaxies, then the surface of the water would remain essentially plane despite its rotation. In any case, it would always be somewhat concave even with the disappearance of the external galaxies due to the rotation of the water relative to the earth, an effect which we are not considering here. In order to take this effect into account we should integrate Eq. (8.14) instead of saying that the force exerted by the earth on the spinning water is simply $\vec{P} = -m_g g \hat{r}$.

This discussion helps to illustrate the striking difference between relational mechanics and the Newtonian or relativistic mechanics. Only in relational mechanics will the curvature of the water will depend on the amount of matter of the distant galaxies. Doubling or halving this amount, while keeping the mass and size of the earth constant, doubles or halves the curvature of the water, assuming the same angular rotation of the water relative to the earth, $\omega_{we} = constant$. In Newtonian mechanics or in Einstein's general theory of relativity, if we are in the universal frame of reference, doubling or halving the mass of the galaxies has no effect on the curvature of the water.

When Mach was criticizing Newton's absolute space he wrote: "Try to fix Newton's bucket and rotate the heaven of fixed stars and then prove the absence of centrifugal forces." Here we have implemented Mach's principle quantitatively, fixing Newton's bucket, rotating the set of fixed stars together with the distant galaxies, and proving the presence of real gravitational centrifugal forces!

9.5   Rotation of the Earth

9.5.1   The Figure of the Earth

We now consider the rotation of the earth relative to the set of fixed stars with a period of one day.

In the frame of the fixed stars the equation of motion of relational mechanics has the same form as in Newton's second law of motion (neglecting the effects of the rotation of the solar system around the center of our galaxy relative to distant galaxies). The changes which appear are: (A) Φm_g instead of m_i; (B) $\vec{a}$ being the acceleration of the earth relative to the fixed stars and not relative to absolute space or to an inertial system of reference; and (C) H_g instead of G. This rotation of the earth will flatten it at the poles. Performing the calculations as in subsection 3.3.2 we find that the ratio of the equatorial radius to the polar radius is given by, with these changes (see Eq. (8.16)):

$$\frac{R_>}{R_<} \approx 1 + \frac{5\Phi\omega_{tf}R_t^3}{4H_g M_t} = 1 + \frac{5\xi}{8}\frac{\omega_{tf}^2}{H_o^2}\frac{\rho_o}{\rho_t} \,, \tag{9.15}$$

where ω_{tf} is the angular rotation of the earth relative to the fixed stars. With the known values of ω_{tf}, H_o, ρ_o, ρ_t and ξ we find that $R_>/R_< \approx 1.004$.

In the earth's frame of reference, and neglecting the translation of the earth around the sun with a period of one year, the equation of motion of relational mechanics takes the form of Eq. (8.47):

$$\sum_{j=1}^{N} \vec{F}_{jm} - \Phi m_g \left[\vec{a}_{me} + \vec{\omega}_{Ue} \times (\vec{\omega}_{Ue} \times \vec{r}_{me}) \right.$$

$$\left. + 2\vec{v}_{me} \times \vec{\omega}_{Ue} + \vec{r}_{me} \times \frac{d\vec{\omega}_{Ue}}{dt} \right] = 0 \,.$$

As the earth is at rest in this frame, $\vec{v}_{me} = 0$ and $\vec{a}_{me} = 0$. Considering also that $d\vec{\omega}_{Ue}/dt = 0$ we arrive at:

$$\sum_{j=1}^{N} \vec{F}_{jm} - m_g\vec{\omega}_{Ue} \times (\vec{\omega}_{Ue} \times \vec{r}_{me}) = 0 \,.$$

In this frame there will appear a real centrifugal force of gravitational origin due to the rotation of distant galaxies around the earth. This centrifugal force flattens the earth at the poles.

What would happen if the external galaxies were annihilated or did not exist? According to relational mechanics the centrifugal force would disappear, except for a small value due to the rotation of the earth relative to the sun, planets and stars belonging to our galaxy. The earth would no longer be flattened, $R_> = R_<$. And Clarke's "absurd" consequence discussed in chapter 5, would be fulfilled. If we double the density of galaxies, then the earth would have a double oblateness ($R_> - R_<$ would double from its present value), provided it kept the same angular rotation relative to the distant universe (with a period of 24 hours).

9.5.2 Foucault's Pendulum

In this case only the Coriolis force will be relevant. The difference as regards Newtonian mechanics is the appearance of Φm_g instead of m_i, H_g instead of G and the rotation of the frame of distant galaxies around the earth instead of the rotation of the earth relative to absolute space. The quantitative explanation will be the same in Newtonian mechanics and in relational mechanics. The final result for the precession of the plane of oscillation of the pendulum will be given by

$$\Omega = -\omega_{ef}\cos\theta = \omega_{ef}\sin\alpha \; , \tag{9.16}$$

where ω_{ef} is the angular rotation of the earth relative to the fixed stars, α is the latitude and $\theta = \pi/2 - \alpha$.

What should be emphasized once again is that relational mechanics offers a physical explanation of the Coriolis force. It is now seen as a real gravitational force due to a relative rotation between the earth and the frame of distant galaxies.

It is worthwhile to present here another calculation of the rotation of the plane of oscillation of Foucault's pendulum. In Figure 9.11 we see the earth centered on O spinning with $\vec{\omega}_{eU} = \omega\hat{z}$ relative to the universal frame of reference U fixed with the distant galaxies. The pendulum is located at a latitude α (*i.e.*, at an spherical angle θ with the SN axis). The weight of the bob m_g is $-m_g g\hat{r}$.

In the earth's frame of reference the equation of motion of relational mechanics is given by Eq. (8.47):

$$\sum_{j=1}^{N} \vec{F}_{jm} - \Phi m_g \left[\vec{a}_{me} + \vec{\omega}_{Ue} \times (\vec{\omega}_{Ue} \times \vec{r}_{me}) \right.$$

$$\left. + 2\vec{v}_{me} \times \vec{\omega}_{Ue} + \vec{r}_{me} \times \frac{d\vec{\omega}_{Ue}}{dt} \right] = 0 \; .$$

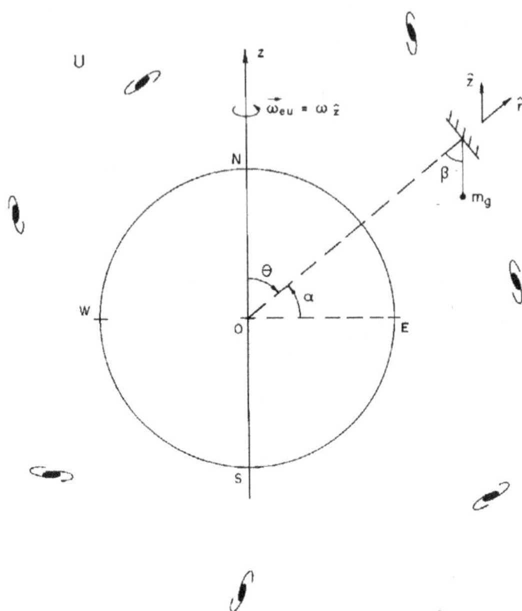

Figure 9.11: Foucault's pendulum in the universal frame U.

As we are in the earth's frame of reference, all velocities, accelerations and angular velocities are relative to the earth's surface. We can consider $d\vec{\omega}_{Ue}/dt = 0$. Moreover, we know that the centrifugal force has no effect in rotating the plane of oscillation of a pendulum, so that we will neglect this force here. The equation of motion for a test particle of gravitational mass m_g can then be written as

$$\sum_{j=1}^{N} \vec{F}_{jm} - 2\Phi m_g \vec{v}_{me} \times \vec{\omega}_{Ue} = \Phi m_g \vec{a}_{me} \ ,$$

where $\vec{a}_{me}$ is the acceleration of the test body relative to the earth and $\vec{\omega}_{Ue}$ is the angular rotation of the distant galaxies (or the fixed stars, neglecting the small acceleration of the solar system relative to the distant galaxies of $\approx 10^{-10}$ m/s^2) relative to the earth. In other words, $\vec{\omega}_{Ue} = -\vec{\omega}_{eU} = -\omega\hat{z}$.

Now, suppose we have a pendulum of length l with a bob of mass m_g oscillating in the gravitational field of the earth. We introduce another coordinate system (x', y', z') or (r', θ', φ'). The origin of this new coordinate system O' is the point of support. We utilize spherical coordinates and consider the z' axis pointing vertically upwards at that location of the earth, orthogonal to the earth's surface at each point, as in Figure 9.12.

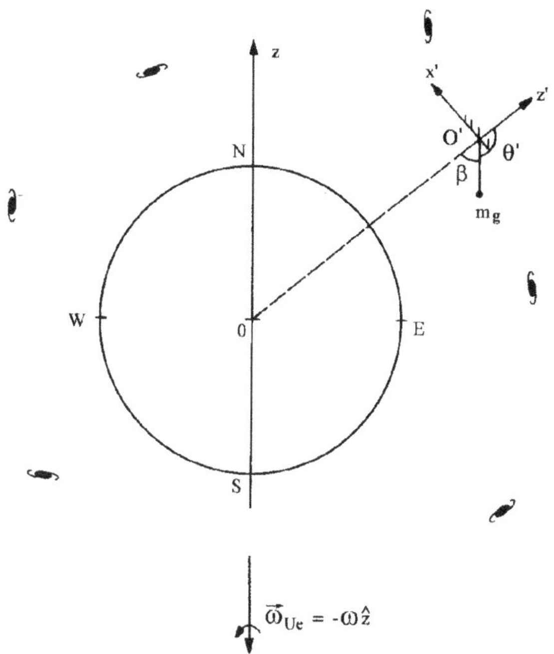

Figure 9.12: Foucault's pendulum in the earth's frame.

The local forces acting on the bob are its weight $\vec{P} = -m_g g\hat{z}'$, with $g = H_g M_t/R_t^2$, and the tension $\vec{T} = -T\hat{r}'$ in the string. Accordingly, the equation of motion of relational mechanics can be written as (with Eq. (8.37 and a similar one for the tension in the string):

$$-m_g\frac{GM_t}{R_t^2}\hat{z}' - T\hat{r}' + m_g\vec{v}_{me}\times\vec{B}_g = m_g\vec{a}_{me}\,.$$

Here we have defined $\vec{B}_g \equiv -2\vec{\omega}_{Ue}$ to let this Coriolis term become similar to Lorentz's magnetic force. It might be called the gravitational magnetic field generated by the rotation of the set of distant galaxies. Now we utilize the fact that $\hat{z}' = \hat{r}'\cos\theta' - \hat{\theta}'\sin\theta'$, $\vec{r}' = l\hat{r}'$, $l = constant$, $\vec{v}_{me} = l\dot{\theta}'\hat{\theta}' + l\dot{\varphi}'\sin\theta'\hat{\varphi}'$, $\vec{a}_{me} = -(l\dot{\theta}'^2 + l\dot{\varphi}'^2\sin\theta'^2)\hat{r}' + (l\ddot{\theta}' - l\dot{\varphi}'^2\sin\theta'\cos\theta')\hat{\theta}' + (l\ddot{\varphi}'\sin\theta' + 2l\dot{\theta}'\dot{\varphi}'\cos\theta')\hat{\varphi}'$. Let us consider that the motion is initially in the $x'z'$ plane. It is then easy to see that a gravitational magnetic field in the y' direction will only change the tension in the string in order to keep its constant length, but will not change the plane of oscillation (observing that the force $m_g\vec{v}\times\vec{B}_g$ will be in this plane). A gravitational magnetic field in the x' direction will not change the plane of oscillation as well. The x' component of the velocity will not be influenced by this field. On the other hand while the bob is going down (with a component of the velocity towards the negative z' direction) the magnetic gravitational force will be in the negative y' direction, while when the bob is going up the force will be in the positive y' direction, both signs reversing when the bob is coming back to the point of release after this half period. This shows that on average the x' component of the gravitational magnetic field does not rotate the plane of oscillation. With the z' component of the gravitational magnetic field the same does not happen. During half a period (while the bob goes down and up) the pendulum experiences a force in the positive y' direction, while during the second half period the force is in the negative y' direction. This clearly rotates the plane of oscillation. From now on we will consider only a gravitational magnetic field in the z' direction, to simplify the analysis: $\vec{B}_g = B_{gz'}\,\hat{z}'$. With this in the $\hat{r}'$ component of the equation of motion we obtain the value of the tension in the string which keeps its length constant: $T = -m_g[(GM_t/R_t^2)\cos\theta' - l\dot{\theta}'^2 - l\dot{\varphi}'^2\sin^2\theta' - B_{gz'}l\dot{\varphi}'\sin^2\theta']$. We are interested only in small oscillations around $\theta' = \pi$. This means that $\sin\theta' \approx \pi - \theta'$ and $\cos\theta' \approx -1$. Utilizing this in the $\hat{\theta}'$ and $\hat{\varphi}'$ components of the equation of motion yields, respectively:

$$\ddot{\theta}' + \left(\frac{GM_t/R_t^2}{l} - B_{gz'}\dot{\varphi}' - \dot{\varphi}'^2\right)\theta' + \pi\left(B_{gz'}\dot{\varphi}' - \frac{GM_t/R_t^2}{l} + \dot{\varphi}'^2\right) = 0\ , \quad (9.17)$$

$$\ddot{\varphi}' = \frac{\dot{\theta}'(B_{gz'} + 2\dot{\varphi}')}{\pi - \theta'}\ . \quad (9.18)$$

We utilize as initial conditions the fact that $\theta' = \theta'_o$, $\dot{\theta}' = 0$, $\varphi' = \varphi'_o$ and $\dot{\varphi}' = 0$. This yields the solution of Eq. (9.18) as:

$$\dot{\varphi}' = \frac{B_{gz'}}{2}\left[1 - \left(\frac{\pi - \theta'_o}{\pi - \theta'}\right)^2\right]\ .$$

This equation shows that if $B_{gz'} = 0$ then $\varphi' = constant$. The solution of Eq. (9.17) satisfying the initial conditions in this case would be: $\theta' = \pi + (\theta'_o - \pi)\cos(\omega_o t)$, where $\omega_o \equiv \sqrt{(GM_t/R_t^2)/l}$.

We now solve Eq. (9.18) iteratively supposing $B_{gz'}^2 \ll \omega_o^2 = g/l$. We put the solution for $B_{gz'} = 0$ in the right hand side of (9.18) and integrate it. The solution of this equation is:

$$\varphi' = \varphi'_o + \frac{B_{gz'}}{2}\left[t - \frac{\tan(\omega'_o t)}{\omega'_o}\right]\ .$$

The period of the pendulum is given by $T = 2\pi/\omega_o$. As $\varphi'(t = 0) = \varphi'_o$ we find that after one period the final value of φ' becomes:

$$\varphi'(t = 2\pi/\omega_o) = \varphi'_o + \frac{\pi B_{gz'}}{\omega'_o} \ .$$

This means that the angular velocity of rotation of the plane of oscillation is found to be given by:

$$\Omega = \frac{\triangle\varphi'}{\triangle t} = \frac{\varphi'(T) - \varphi'_o}{T} = \frac{B_{gz'}}{2} \ .$$

Now let us look at Figure 9.12.

The stars and distant galaxies rotate in the north-south direction relative to the earth. This is the direction of $\vec{\omega}_{Ue} = -\omega\hat{z}$. For a pendulum oscillating at latitude α (at Paris, for instance, where $\alpha = 48^o \ 51'$ N) the z' component of the gravitational magnetic field will be given by $B_{gz'} = B_g \cos\theta = B_g \cos(90^o - \alpha) = B_g \sin\alpha = 2\omega_{Ue} \sin\alpha$, where we utilized the definition introduced earlier, namely: $\vec{B}_g = -2\vec{\omega}_{Ue}$. This means that the angular rotation of the plane of oscillation will be given by:

$$\Omega = \frac{B_{gz'}}{2} = \omega_{Ue} \sin\alpha \ .$$

This is the observed value of the rotation of the plane of oscillation. The approximation utilized in this calculation that $B_{gz'}^2 \ll \omega_o^2 = (GM_t/R_t^2)/l$ is easily justified observing that in Foucault's real experiment we had $l = 11 \ m$ such that $\omega_o \approx 1 \ rad/s \gg B_{gz'} = 2\omega_{Ue} \sin\alpha = 2(7.3 \times 10^{-5} \ \text{rad/s}) \times 0.75 = 10^{-4} \ \text{rad/s}$.

In relational mechanics we can interpret this rotation as an analog of the magnetic Lorentz force law. In this case the spinning set of distant galaxies generates a gravitational magnetic field $\vec{B}_g = -2\vec{\omega}_{Ue}$ which exerts a force on moving masses given by $m_g\vec{v}_{me} \times \vec{B}_g$.

The flattened figure of the earth or Foucault's pendulum can no longer be utilized as proofs of the earth's real rotation. In relational mechanics, both facts can be equally explained with the frame of distant galaxies at rest (exerting a gravitational force $-\Phi m_g\vec{a}_{mU}$ on bodies at the earth's surface) while the earth rotates relative to this frame, or with the earth at rest while the distant galaxies rotate around it exerting a gravitational force $-\Phi m_g(\vec{a}_{me} + 2\vec{v}_{me} \times \vec{\omega}_{Ue} + \vec{\omega}_{Ue} \times (\vec{\omega}_{Ue} \times \vec{r}_{me}))$ on bodies at the earth's surface. Both explanations are equally correct and yield the same effects. It then becomes a matter of convenience or of convention to choose the earth, the distant galaxies or any other body or frame of reference to be considered at rest. This is an important and deep result of relational mechanics, which had not been implemented by any other formulation of mechanics up to now.

We then acquire a new comprehension of Foucault's pendulum. Let us present it only in the simplest case of a pendulum oscillating over the North pole. We conclude that the plane of oscillation is fixed relative to the frame of distant galaxies, no matter what the rotation of the earth relative to this frame, as in Figure 9.13. In this Figure we are in the universal frame of reference U watching the earth rotate below us. While the earth rotates, the plane of oscillation remains fixed relative to U.

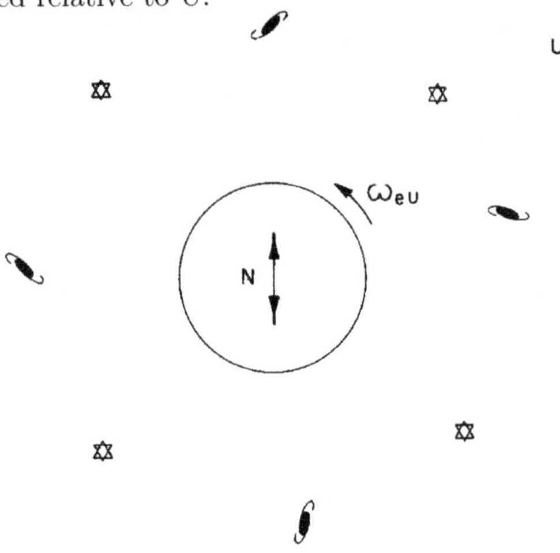

Figure 9.13: Plane of oscillation of Foucault's pendulum fixed relative to the set of distant galaxies.

For an observer fixed on the earth, the explanation is the same. Accordingly, he can say that the distant galaxies rotating around the earth make the plane of oscillation of the pendulum rotate with them, as in Figure 9.14: $\vec{\Omega}_{pe} = \vec{\omega}_{Ue}$.

If all the matter in the universe were annihilated, except the pendulum and the earth, then according to relational mechanics the plane of oscillation of the pendulum would be fixed relative to the earth. The reason for this is that the Coriolis force is due, in relational mechanics, to an interaction between the test body and the distant galaxies. If these galaxies disappear, the Coriolis force disappears as well, the same happening with $-\Phi m_g \vec{a}_{mU}$. If the density of external galaxies were multiplied by a constant value k, keeping the earth and the bob with the same gravitational masses, then the bob would behave as having a Newtonian inertial mass given by km_g. Utilizing this, we can com-

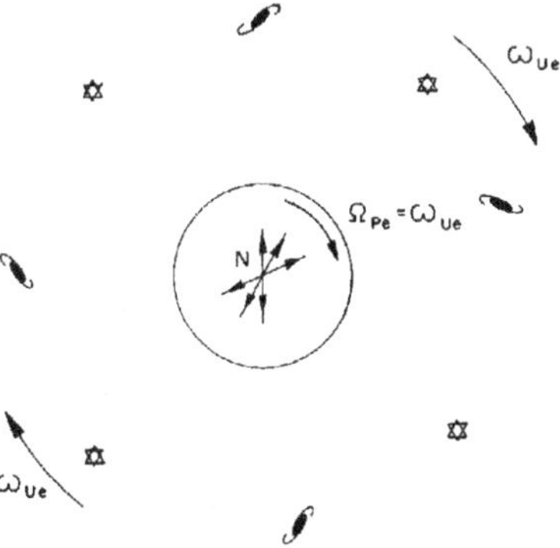

Figure 9.14: Rotating universe making the plane of oscillation rotate with it.

pare the results of relational mechanics with those of Newtonian or relativistic mechanics.

Chapter 10

Beyond Newton

In the previous chapters we saw how we can recover results analogous to the results of Newtonian mechanics with relational mechanics. Moreover, we were able to explain many puzzles of classical physics, such as the proportionality between inertial and gravitational masses, the origin of the centrifugal and Coriolis forces, *etc.*

In this chapter we discuss some phenomena which are beyond the Newtonian theory. These will come from the extra terms which appear in Weber's force applied to gravitation, compared with Newton's inverse square law.

10.1 Precession of the Perihelion of the Planets

We begin by discussing the problem of two bodies moving under gravitational interaction in the presence of distant galaxies. In the case of the solar system we can apply Eq. (8.51) so that the equation of motion for body 1 in the frame of the fixed stars takes the form

$$\vec{F}_{21} - \Phi m_{g1}\vec{a}_{1f} = 0 \ .$$

Here $\vec{F}_{21}$ is the force exerted by 2 on 1 and $\vec{a}_{1f}$ is the acceleration of body 1 relative to the fixed stars (which will be essentially the same as its acceleration relative to distant galaxies). Analogously, for body 2 we obtain

$$\vec{F}_{12} - \Phi m_{g2}\vec{a}_{2f} = 0 \ .$$

We consider here the sun interacting with a planet. We can assume the planets to be material points, as their diameters are much smaller than their distances to the sun. In this problem, the sun can also be considered to be a material point. As a matter of fact, the force exerted by the sun of radius R_s on an external material point 1 is obtained integrating Eq. (8.14). As we have shown before [132], the terms multiplying the second ξ in this equation are at least 6×10^{-4} smaller than those multiplying the first ξ for the planetary system. This means that we can treat the sun as a material point in this problem.

The force exerted by the sun 2 on a planet 1 according to Weber's expression is then given by Eq. (8.4):

$$\vec{F}_{21} = -H_g m_{g1} m_{g2} \frac{\hat{r}_{12}}{r_{12}^2} \left[1 - \frac{\xi}{c^2} \left(\frac{\dot{r}_{12}^2}{2} - r_{12} \ddot{r}_{12} \right) \right] = -\vec{F}_{12} \ .$$

This means that the equations of motion take the form (with Eq. (8.37):

$$m_{g1} \vec{a}_{1f} = - G m_{g1} m_{g2} \frac{\hat{r}_{12}}{r_{12}^2} \left[1 - \frac{\xi}{c^2} \left(\frac{\dot{r}_{12}^2}{2} - r_{12} \ddot{r}_{12} \right) \right] \ , \qquad (10.1)$$

$$m_{g1} \vec{a}_{2f} = + G m_{g1} m_{g2} \frac{\hat{r}_{12}}{r_{12}^2} \left[1 - \frac{\xi}{c^2} \left(\frac{\dot{r}_{12}^2}{2} - r_{12} \ddot{r}_{12} \right) \right] \ . \qquad (10.2)$$

Adding these two equations yields the conservation of the total linear momentum of the system sun-planet relative to the fixed stars:

$$m_{g1} \vec{a}_{1f} + m_{g2} \vec{a}_{2f} = \frac{d}{dt} (m_{g1} \vec{v}_{1f} + m_{g2} \vec{v}_{2f}) = 0 \ .$$

The center of mass of the sun-planet system, $\vec{R} \equiv (m_{g1} \vec{r}_{1f} + m_{g2} \vec{r}_{2f})/(m_{g1} + m_{g2})$ then moves with a constant velocity relative to the fixed stars.

The difference between the accelerations in Eqs. (10.1) and (10.2) yields:

$$\vec{a}_{12} \equiv \vec{a}_{1f} - \vec{a}_{2f} = -G(m_{g1} + m_{g2}) \frac{\hat{r}_{12}}{r_{12}^2} \left[1 + \frac{\xi}{c^2} \left(r_{12} \ddot{r}_{12} - \frac{\dot{r}_{12}^2}{2} \right) \right] \ . \qquad (10.3)$$

This shows that $\vec{a}_{12}$ is parallel to $\hat{r}_{12}$.

We can define a relative angular momentum in the frame of fixed stars by

$$\vec{L}_{12} \equiv \vec{r}_{12} \times (M_g \vec{v}_{12}) \ ,$$

where $M_g \equiv m_{g1} + m_{g2}$ is the total gravitational mass of the sun-planet system. Taking the derivative of this equation with respect to time and utilizing the fact that $\vec{v}_{12} \times \vec{v}_{12} = 0$ and the previous result that $\vec{a}_{12}$ is parallel to $\hat{r}_{12}$ yields a zero value. This means that $\vec{L}_{12}$ is a constant in time. Moreover, $\vec{r}_{12}$ and $\vec{v}_{12}$ lie in a plane whose normal is parallel to $\vec{L}_{12}$. We then choose a coordinate system centered on the sun such that its z axis is parallel to $\vec{L}_{12}$. In this frame of reference the planet will always move in the xy plane. Writing Eq. (10.3) in plane polar coordinates yields two equations, one for the $\hat{\varphi}$ component and another for the $\hat{\rho}$ component, namely

$$\rho\ddot{\varphi} + 2\dot{\rho}\dot{\varphi} = 0 \, , \tag{10.4}$$

$$\ddot{\rho} - \rho\dot{\varphi}^2 = -GM_g \left[\frac{1}{\rho^2} + \frac{\xi}{c^2} \left(\frac{\ddot{\rho}}{\rho} - \frac{\dot{\rho}^2}{2\rho^2} \right) \right] \, . \tag{10.5}$$

The first of these equations yields the conservation of angular momentum. This means that the quantity $H \equiv \rho^2\dot{\varphi}$ is a constant for all time.

Defining $u \equiv 1/\rho$ and utilizing a standard prescription, the second equation can be put in the form

$$\frac{d^2 u}{d\varphi^2} + u = GM_g \left\{ \frac{1}{H^2} - \frac{\xi}{c^2} \left[\frac{1}{2} \left(\frac{du}{d\varphi} \right)^2 + u\frac{d^2 u}{d\varphi^2} \right] \right\} \, . \tag{10.6}$$

The exact solution of this equation in terms of elliptic functions can be found in previous works [180] and [12]. Here we solve it iteratively following another study [132]. Observing that the second and third terms in the square bracket are much smaller than the first one, we seek a solution in the form $u(\varphi) = u_o(\varphi) + u_1(\varphi)$, with $|u_o| \gg |u_1|$, and where u_o and u_1 satisfy the equations

$$\frac{d^2 u_o}{d\varphi^2} + u_o = \frac{GM_g}{H^2} \, ,$$

$$\frac{d^2 u_1}{d\varphi^2} + u_1 = -GM_g\frac{\xi}{c^2} \left[\frac{1}{2} \left(\frac{du_o}{d\varphi} \right)^2 + u_o\frac{d^2 u_o}{d\varphi^2} \right] \, .$$

The solution of the first equation is the classical result

$$u_o(\varphi) = \frac{GM_g}{H^2} + A\cos(\varphi - \varphi_o) \, ,$$

where A and φ_o come from the initial conditions. With this solution for u_o the solution for u_1 is found to be

$$u_1(\varphi) = \frac{G^2 M_g^2 A}{2H^2} \frac{\xi}{c^2} (\varphi - \varphi_o)\sin(\varphi - \varphi_o) + \frac{GM_g A^2}{2} \frac{\xi}{c^2} \sin^2(\varphi - \varphi_o) \, .$$

The turning points, at which the distance of the planet from the sun is a maximum or a minimum, are given by $du/d\varphi = 0$. We can see from these equations that $\varphi = \varphi_o$ is one solution. After one revolution, the turning point will be near $\varphi_o + 2\pi$. Expanding $du/d\varphi$ around this value and equating to zero yields

$$\varphi \approx \varphi_o + 2\pi + \frac{\pi G^2 M_g^2}{H^2} \frac{\xi}{c^2} \; .$$

The advance of the perihelion in one revolution is then given by

$$\triangle\varphi = \pi \frac{\xi}{c^2} \frac{G^2 M_g^2}{H^2} = \pi \frac{\xi}{c^2} \frac{GM_g}{a(1 - \varepsilon^2)} \; , \tag{10.7}$$

where a is the semimajor axis and ε is the eccentricity of the orbit. With the value of $\xi = 6$ we arrive at a result which is well observed in the solar system and which agrees algebraically with the one given by Einstein's general theory of relativity.

Despite this coincidence, the orbit equation obtained in general relativity is given by

$$\frac{d^2 u}{d\varphi^2} + u = \frac{GM}{H^2} + \frac{3GM}{c^2} u^2 \; .$$

Comparison of this equation with (10.6) shows that they are not equivalent in general. At zeroeth order both yield the ellipses, parabolas and hyperbolas of Newtonian theory. At first order both yield the same precession of the perihelion of the planets. At the second order they differ from one another. At present, we cannot distinguish the second order terms of these models utilizing the data from the solar system.

Before comparing these two equations in second order, it would be more important to review the calculations of the precession of the perihelion of the planets utilizing these two theories, but taking into account the perturbation due to other planets. As is well known, the Newtonian theory explains most of the observed precession of the perihelion of the planets taking into account the perturbations due to other planets. But there remains a small residual value which the Newtonian theory cannot explain. It is this residual value which is explained in general relativity and in relational mechanics by Eq. (10.7). In order to be coherent, it would be better to calculate the precession due to the perturbation of other planets again, not with the Newtonian inverse square force, but with general relativity and Weber's force applied to gravitation. We then see what residual values remain with both theories (they may be different from one another, or from the one given by the Newtonian theory). After this calculation we can compare the residual values which cannot be explained in both models considering only the influence of the other planets, if they still exist, with Eq. (10.7). In other words, taking into account the effect due to the sun.

10.2 Anisotropy of Inertial Mass

We discuss here an important consequence of any model that seeks to implement Mach's principle. We concentrate our analysis on Weber's law applied to gravitation. When we identify relational mechanics with Newtonian mechanics we conclude that the inertia of a body, its inertial mass, is due to a gravitational interaction with distant masses. One consequence of this fact is as follows: if this distribution is anisotropic, the effective inertial mass of the test body should be anisotropic as well. We illustrate this effect by analysing the "two-body problem" (two bodies plus the distant galaxies).

In the previous section we found that the equation of motion for a planet m_{g1} interacting with the sun m_{g2} and the distant galaxies in relational mechanics is given by Eq. (10.1):

$$m_{g1}\vec{a}_{1f} = -\, Gm_{g1}m_{g2}\frac{\hat{r}_{12}}{r_{12}^2}\left[1 - \frac{\xi}{c^2}\left(\frac{\dot{r}_{12}^2}{2} - r_{12}\ddot{r}_{12}\right)\right] \ .$$

In the approximation in which $m_{g2} \gg m_{g1}$, we can disregard the motion of the sun relative to the frame of fixed stars, and consider this essentially as a "one body problem" under the influence of a central force. For the motion in the xy plane centered on the sun we have, with cylindrical coordinates: $\vec{a}_{1f} = (\ddot{\rho} - \rho\dot{\varphi}^2)\hat{\rho} + (\rho\ddot{\varphi} + 2\dot{\rho}\varphi)\hat{\varphi}$, $r_{12} = \rho$, $\dot{r}_{12} = \dot{\rho}$, $\ddot{r}_{12} = \ddot{\rho}$, $\hat{r}_{12} = -\rho\hat{\rho}$, where ρ is the distance of the planet to the sun. The $\dot{\varphi}$ component of this equation yields the conservation of angular momentum, Eq. (10.4). The radial component is given by Eq. (10.5), which can be written in this approximation where $m_{g2} \gg m_{g1}$, as:

$$m_r\ddot{\rho} - m_t\rho\dot{\varphi}^2 = -G\frac{m_{g1}m_{g2}}{\rho^2} + Gm_{g1}m_{g2}\frac{\xi}{c^2}\frac{\dot{\rho}^2}{2\rho^2} \ ,$$

where $m_t \equiv m_{g1}$ is the usual gravitational mass of the planet and $m_r \equiv m_{g1}(1 + Gm_{g2}\xi/\rho c^2)$. Consequently, apart from the second term on the right hand side, this equation is analogous to the Newtonian equation of motion with an effective radial inertial mass m_r and an effective tangential inertial mass m_t. These two masses, m_r and m_t, are different from one another due to the fact that the sun is interacting with the planet along the radial direction connecting them, but not along the tangential direction. It is exactly this term in m_r which will be responsible for the precession of the perihelion of the planets, the second term on the right hand side of this equation yielding no precession.

We may then consider the precession of the perihelion of the planets as another strong fact supporting (although not proving) Mach's principle and the anisotropy of the inertial mass of bodies. This seems to have been seen clearly for the first time by Schrödinger [125] (English translation: [83]). He calculated the precession of the perihelion of the planets utilizing a potential energy analogous to Weber's, instead of working with the forces. The interaction gravitational energy between the planet and the distant galaxies has been found to be given by $m_g v^2/2$, where m_g is the gravitational mass of the planet and v its velocity relative to the universal frame of reference. Utilizing polar coordinates in the plane of the motion and the approximation that the mass of the sun is much greater than the mass of the planet ($M_g \gg m_g$) yields $v^2 = \dot{\rho}^2 + \rho^2\dot{\varphi}^2$. The energy of interaction between the planet and the sun is given by Eq. (8.3). So the total constant energy for the planet is found to be (with Eq. (8.37)):

$$m_g \frac{\dot{\rho}^2 + \rho^2\dot{\varphi}^2}{2} - G\frac{M_g m_g}{\rho}\left(1 - \xi\frac{\dot{\rho}^2}{2c^2}\right) = constant \ .$$

This equation can be written as:

$$\frac{m_r}{2}\dot{\rho}^2 + \frac{m_t}{2}\rho^2\dot{\varphi}^2 - G\frac{M_g m_g}{\rho} = constant \ ,$$

where $m_t \equiv m_g$ and $m_r \equiv m_g(1 + \xi G M_g/\rho c^2)$. Hence, the law for the conservation of energy becomes analogous to the Newtonian law, provided there are different radial and tangential effective inertial masses. This was the conclusion of Schrödinger when he wrote (see [83, especially p. 151]): "The presence of the sun has, in addition to the gravitational attraction, also the effect that the planet has a somewhat greater inertial mass 'radially' than 'tangentially'." The different effective masses in the radial and tangential components yield the precession of the perihelion of the planets in relational mechanics. The observation of the precession of the perihelion of the planets can then be regarded as a proof of the anisotropy of inertial mass. Schrödinger goes on to conclude that the inertia of a body should be greater in the galactic plane than perpendicular to it [*ibidem*, p. 153]: "A mass distribution like that established for the radiating stars would have to have the consequence that bodies are subject to a greater inertial resistance in the galactic plane as at right angles to it."

As we have seen, the effective inertial mass of a body is different when there are anisotropies in the distribution of matter around the body. This is observed in purely gravitational interactions, as in the motions of the solar system. In electromagnetic interactions, this effect also appears. Recently we showed that the self-induction of an electrical circuit can be derived from the component of Weber's electromagnetic force which depends on the acceleration of the test charges [181]. We found that the self-induction is a measure of the effective inertial mass of the mobile electrons. We also derive the fact that in a conducting cylindrical shell of length ℓ and radius a the self-inductance for the case of a longitudinal current (current along the axial direction) will be different from the the self-inductance in the case of a poloidal current (current in the

φ direction of polar coordinates). This shows the anisotropy in the effective inertial mass discussed here. After all, it is a known experimental fact that the self-induction for the same cylindrical shell is different for different directions of current flow. As this effect can be derived from Weber's electrodynamics, it yields greater support to the idea of an anisotropic effective inertial mass.

The effect should also appear in other electromagnetic situations. If a test charge is interacting with fixed anisotropic distributions of charge around it, it should behave according to Weber's electrodynamics as if it had an effective inertial mass which depends on the geometry of the problem, on the direction of motion and on the electrostatic potential energy where it is located ([9], [10], [12, Section 7.2], [14] and [15]). Experimental tests of this fact, which does not appear in Maxwell-Lorentz electrodynamics, have been proposed in other studies ([17] and [18]). We believe Weber's electrodynamics will be vindicated by these experiments. In order to perform the test, it is important to keep the anisotropic distributions of charge, which are acting on the test charge, fixed relative to one another and to the laboratory, while the test charge is accelerated relative to them and to the laboratory. The experiment cannot be performed by charging a Faraday cage and accelerating charges inside it. The reason for this is that in this latter situation there are free charges in the metallic Faraday cage which will move when the test charge is accelerated inside it, responding to the motion of the test charge. This will mask the effect to be observed (the possible change in the effective inertial mass of the test charge). To perform the experiment it is important to charge a dielectric, which will keep the net charges fixed relative to it no matter what the motion of the test charge inside it.

For a different approach on this topic see Eby's work [182].

We conclude here by calling attention to the work of the Nobel prize winner Maurice Allais and of other writers quoted in his work. In optical experiments and in experiments performed with pendulums they found anomalous effects which might be interpreted as an anisotropy in the inertial mass of the test particle connected with the astronomical bodies (Moon, *etc.*) [183].

10.3 High Velocity Particles

There are some indications that the correct expression for the kinetic energy of a test particle is given by $mc^2(1/\sqrt{1 - v^2/c^2} - 1)$ instead of $mv^2/2$. These indications come from experiments with high speed electrons and in high energy collisions of charged particles. Following this clue, Schrödinger in his important paper of 1925, and independently Wesley in 1990, proposed a modification of Weber's potential energy for gravitation ([125], [184] and [12, Section 7.7]). What they proposed was a gravitational potential energy between two gravitational masses m_{g1} and m_{g2} given by

$$U_{12} = \beta \frac{m_{g1}m_{g2}}{r_{12}} + \gamma \frac{m_{g1}m_{g2}}{r_{12}} \frac{1}{(1 - \dot{r}_{12}^2/c^2)^{3/2}} \ . \tag{10.8}$$

Schrödinger proposed $\beta = -3G$ and $\gamma = 2G$, while Wesley took $\beta = -4G/3$ and $\gamma = G/3$. When $\dot{r}_{12} = 0$ we recover the Newtonian potential energy. Expanding this expression up to second order in $\dot{r}_{12}/c$ yields a potential energy for gravitation analogous to Weber's.

The force exerted by 2 on 1 is obtained by $\vec{F}_{21} = -\hat{r}_{12}dU_{12}/dr_{12}$ or by $dU_{12}/dt = -\vec{v}_{12} \cdot \vec{F}_{21}$. This yields:

$$\vec{F}_{21} = m_{g1}m_{g2}\frac{\hat{r}_{12}}{r_{12}^2}\left[\beta + \gamma\left(1 - \frac{\dot{r}_{12}^2}{c^2} - 3\frac{r_{12}\ddot{r}_{12}}{c^2}\right)\left(1 - \frac{\dot{r}_{12}^2}{c^2}\right)^{-5/2}\right] \ . \tag{10.9}$$

We now integrate both expressions for a test particle of gravitational mass m_{g1} interacting with the isotropic distribution of mass around it (with the isotropic distribution of distant galaxies with a constant gravitational mass density ρ_o). We perform the integration in the universal frame of reference U, frame in which the set of distant galaxies is essentially at rest and without rotation. The velocity and acceleration of m_{g1} relative to this frame are given by, respectively: $\vec{v}_{1U}$ and $\vec{a}_{1U}$. The result of integration up to the Hubble distance c/H_o is

$$U_{Im} = 2\pi\frac{m_{g1}\rho_o c^2}{H_o^2}\left(\beta + \frac{\gamma}{\sqrt{1 - v_{1U}^2/c^2}}\right) \ , \tag{10.10}$$

$$\vec{F}_{Im} = -2\pi\gamma\frac{\rho_o}{H_o^2}\left[\frac{m_{g1}\vec{a}_{1U}}{\sqrt{1 - v_{1U}^2/c^2}} + \frac{m_{g1}\vec{v}_{1U}(\vec{v}_{1U} \cdot \vec{a}_{1U})}{c^2(1 - v_{1U}^2/c^2)^{3/2}}\right]$$

$$= -2\pi\gamma\frac{\rho_o}{H_o^2}\frac{d}{dt}\left(\frac{m_{g1}\vec{v}_{1U}}{\sqrt{1 - v_{1U}^2/c^2}}\right) \ . \tag{10.11}$$

If we wanted to integrate to infinity, it would only be necessary to include an exponential decay in both terms on the right hand side of Eq. (10.8).

With the principle of dynamical equilibrium, it is then possible to derive a relativistic kinetic energy analogous to the Einstenian energy, and an equation of motion analogous to his equation of motion. But despite the similarity in the form of the equations, there are many differences in both models. The first is that results (10.10) and (10.11) were obtained after a gravitational interaction of the test body with the distant masses in the cosmos in relational mechanics, while this is not the case in Einstein's theory of relativity. As a consequence, the masses which appear in (10.10) and (10.11) are gravitational masses, while the Einsteinian masses are inertial masses in the Newtonian sense, with inertia related to space and not to distant matter. Moreover, the velocities and accelerations of the test body which appear in these equations are relative to the distant universe in relational mechanics, while in Einstein's theory they are relative to an arbitrary inertial frame of reference.

Let us consider two bodies 1 and 2 interacting with one another and with the distant galaxies. The principle of dynamical equilibrium applied to Eqs. (10.9) and (10.11) yields the equation of motion for body 1 in the universal frame of reference, namely:

$$m_{g1}m_{g2}\frac{\hat{r}_{12}}{r_{12}^2}\left[\beta + \gamma\left(1 - \frac{\dot{r}_{12}^2}{c^2} - 3\frac{r_{12}\ddot{r}_{12}}{c^2}\right)\left(1 - \frac{\dot{r}_{12}^2}{c^2}\right)^{-5/2}\right]$$

$$- 2\pi\gamma\frac{\rho_o}{H_o^2}\left[\frac{m_{g1}\vec{a}_{1U}}{\sqrt{1 - v_{1U}^2/c^2}} + \frac{m_{g1}\vec{v}_{1U}(\vec{v}_{1U} \cdot \vec{a}_{1U})}{c^2(1 - v_{1U}^2/c^2)^{3/2}}\right] = 0 \ .$$

Assuming that bodies 1 and 2 are orbiting around one another in this frame, with $\dot{r}_{12} = 0$ and $\ddot{r}_{12} = 0$, and assuming that $v_{1U}^2 \ll c^2$, this equation reduces to:

$$\frac{(\beta + \gamma)H_o^2}{2\pi\gamma\rho_o}m_{g1}m_{g2}\frac{\hat{r}_{12}}{r_{12}^2} = m_{g1}\vec{a}_{1U} \ .$$

This shows that we can recover Newtonian mechanics only if the following relation is exactly valid:

$$\frac{(\beta + \gamma)H_o^2}{2\pi\gamma\rho_o} = -G \ .$$

Utilizing the values of β and γ given by Schrödinger and Wesley, and the observational values of H_o, ρ_o and G, we find that this relation is approximately valid. We cannot say that this relation is exactly valid, due to uncertainties in the observational values of H_o and ρ_o. In any event, we see once more that with $\beta/\gamma \approx -1$ we obtain as a consequence of relational mechanics that $H_o^2/\rho_o \approx G$, a result which is confirmed by the observational values of these independent quantities.

An important topic in relational mechanics deals with the interaction between matter and radiation: deflection of light in a gravitational field and gravitational redshift. The first paper dealing with the bending of light utilizing a Weber's type law for gravitation has been given by Ragusa [185]. In order to obtain the correct bending of light and the correct precession of the perihelion of the planets he introduced two parameters, one in front of $\dot{r}^2$ and another in front of $r\ddot{r}$. However, although it worked correctly, this solution has problems with the conservation of energy as has been pointed out by Bunchaft and Carneiro [186]. But as they say in the paper, if there are in the gravitational law terms of order higher than $1/c^2$, they would not affect the calculations for the precession of the perihelion (low velocity phenomenon) but would affect the calculation for the gravitational deflection of light. Once more we need further research in this direction before drawing final conclusions. To our knowledge there are not yet publications with complete calculations for the gravitational redshift performed with Weber's law for gravitation and with generalizations of it to high velocity particles (for velocities close to c or equal to c).

10.4   Experimental Tests of Relational Mechanics

In principle we might test a Weber's force applied to gravitation by utilizing the result (8.12), yielding the force of a spherical shell of radius R on an internal test particle:

$$d\vec{F}_{Mm}(r < R) = -\frac{4\pi}{3} H_g \frac{\xi}{c^2} m_g \rho_g R dR \left[\vec{a}_{mS} + \vec{\omega}_{MS} \times (\vec{\omega}_{MS} \times \vec{r}_{mS}) \right.$$

$$\left. + 2\vec{v}_{mS} \times \vec{\omega}_{MS} + \vec{r}_{mS} \times \frac{d\vec{\omega}_{MS}}{dt} \right] .$$

Now suppose we are in the earth's frame of reference in such a situation that relation (8.48) is satisfied. We then surround a test particle of gravitational mass m_g by the previous spherical shell of mass $dM_g = 4\pi\rho_g R^2 dR$ at rest relative to the earth, while the test body is accelerated relative to it by other forces. The force exerted by this shell on m_g is then given by $-H_g \xi m_g M_g \vec{a}_{me}/3Rc^2$. The gravitational force exerted by the distant galaxies on m_g in the absence of the spherical shell has been found in the approximation of (8.48) to be given by $-\Phi m_g \vec{a}_{me}$. The equation of motion of relational mechanics then takes the form:

$$\sum_{j=1}^{N} \vec{F}_{jm} - \frac{H_g \xi m_g M_g}{3Rc^2} \vec{a}_{me} - \Phi m_g \vec{a}_{me} = 0 .$$

That is,

$$\sum_{j=1}^{N} \vec{F}_{jm} = \Phi m^* \vec{a}_{me} \; .$$

Here m^* can be considered the effective inertial mass of the test body surrounded by the spherical shell. It is given by (with Eqs. (8.16) and (8.37):

$$m^* \equiv m_g \left(1 + \frac{G\xi M_g}{3Rc^2} \right) \; .$$

In principle this increase in the effective inertial mass of the test body surrounded by the spherical shell might be tested experimentally. We should first study the acceleration of a test body not surrounded by the spherical shell, for instance, by accelerating a charged particle with other charges, currents and magnets; or letting a body fall freely on the surface of the earth. Then we surround the test body with the neutral shell and accelerate the test body again with the same bodies (other charges, currents and magnets from the previous example; or the earth in the free fall experiment) and study its new acceleration, which should be different now. The new acceleration or any effect depending on this acceleration, such as the radius of curvature of the curved orbit of the test body should have been changed, according to relational mechanics. The problem is the small value of this increase. The percentage increase is given by

$$\frac{m^* - m_g}{m_g} = \frac{G\xi M_g}{3Rc^2} \; .$$

If we have a spherical shell of mass 100 kg, radius 1 m and take $\xi = 6$ this yields $\approx 10^{-25}$, which is obviously undetectable.

It should be observed that nothing of this would happen in Newtonian mechanics or in Einstein's general theory of relativity. The reason is that in these theories there are no effects due to a spherically symmetric and stationary distribution of masses around a test body.

An analogous test for Weber's law applied to electromagnetism has been proposed in [18]. The idea is to compare Weber's electrodynamics and Lorentz's force law. Here we analyse the motion of a test charge inside and outside a charged spherical shell. For details see the paper and [12, Section 9.3: Charged Spherical Shell].

If we had surrounded the test body by an anisotropic distribution of mass (a hollow cube or a hollow cylinder) the effective inertial mass would be different in different directions. The next step after testing the previous effect would be to check this prediction.

It might also be possible to test this anisotropy in the effective inertial mass of test bodies by taking into account existing anisotropies in the distribution of matter in nature. As we have seen, the precession of the perihelion of the planets may be viewed, from this standpoint, as due to this anisotropy owing to the sun's influence radially but not tangentially. If the test body is near the surface of the earth, there is the anisotropy due to the proximity of the earth. The effective inertial mass of a test body moving vertically should be different from the effective inertial mass of the same test body moving horizontally relative to the earth's surface. By the same token, the inertia of a body being accelerated in the direction of the moon, or of the sun, or of the center of our galaxy, should be different from the inertia of the same body being accelerated in a plane orthogonal to these directions. The effect can be estimated by looking at Weber's force applied to gravitation, Eq. (8.4):

$$\vec{F}_{21} = -H_g m_{g1} m_{g2} \frac{\hat{r}_{12}}{r_{12}^2}\left[1 - \frac{\xi}{c^2}\left(\frac{\dot{r}_{12}^2}{2} - r_{12}\ddot{r}_{12}\right)\right] .$$

If the effective inertial mass of 1 when moving in a plane orthogonal to the straight line connecting it to 2 is $m_{it} = m_{g1}$, this equation shows that its effective inertial mass when accelerated in the direction of 2 will be given by the following order of magnitude: $m_{ir} = m_{g1}[1 + \xi G m_{g2}/r_{12}c^2]$. The percentage change is then given by:

$$\frac{m_{ir} - m_{it}}{m_{it}} \approx \frac{\xi G m_{g2}}{r_{12}c^2} .$$

In this analysis we will suppose $\xi = 6$. Taking $m_{g2} = 3 \times 10^{41}$ kg as the mass of our galaxy and $r_{12} = 2.5 \times 10^{20}$ m as the distance of the solar system to the center of our galaxy yields: 5×10^{-6}. There should be a difference of one part in 10^6 comparing the inertia of a planet or any other body accelerated in the direction of the center of our galaxy and accelerated normal to this direction. Taking $m_{g2} = 2 \times 10^{30}$ kg as the sun's mass and $r_{12} = 1.5 \times 10^{11}$ m as our distance to the sun yields 2×10^{-7}. Taking $m_{g2} = 7 \times 10^{22}$ kg as the moon's mass and $r_{12} = 3.8 \times 10^8$ m as our distance to the moon yields 8×10^{-13}. Taking $m_{g2} = 6 \times 10^{24}$ kg as the earth's mass and $r_{12} = 6 \times 10^6$ m as the earth's radius yields 5×10^{-9}. This shows that we could observe this effect by performing experiments in which the test body moves vertically or horizontally relative to the earth's surface. To this end, the precision should be of the order of 10^{-10}. If we want to compare the anisotropy due to our own galaxy, the precision needs to be 10^{-7}.

To estimate these effects, we are supposing a purely gravitational experiment. Moreover, we are neglecting the influence of the term in $\dot{r}_{12}^2/c^2$ which might mask the effect being looked for. A careful analysis and calculation should be performed in each specific case before reaching any general conclusion.

In an interesting paper published in 1958, Cocconi and Salpeter predicted these ideas by considering a general implementation of Mach's principle, not necessarily connected with Weber's force [187]. They do not mention Weber's force or Schrödinger's work, but only Mach's ideas. In any case, Weber's force fits nicely in their general approach. After all, they considered the possibility that the contribution to the inertia of a test body of mass m resulting from its interaction with a mass M separated by a distance r has the following properties: proportional to M, falls as r^ν and depends on the angle θ between the acceleration of the test body and the straight line connecting them. Weber's component, which yields the inertia of a body, has these properties with $\nu = 1$, as it is of the form $-Gm_{g1}M_{g2}\hat{r}_{12}\ddot{r}_{12}/r_{12}c^2$. Motivated by this paper, many experiments were devised to find this anisotropy in the inertial mass of the bodies ([188], [189], [190], [191], *etc.*) They looked for anisotropies utilizing the Zeeman splitting in an atom, the Mössbauer effect, nuclear magnetic ressonance, *etc.* All of these yielded a null result.

How can we explain their negative findings in the context of relational mechanics? The first answer was given by Dicke [192], who observed that according to Mach's principle this effect must be there, but it should be observed that this anisotropy of the inertial mass is universal, the same for all particles (including photons and pions). Due to this universality of the anisotropy, it would be unobservable locally. The second answer was given by Edwards [193], who observed that the effect of such an anisotropy on local measuring instruments must be carefully considered before one can draw the conclusion that the anisotropy of the inertial mass has been ruled out by these experiments. We agree with Dicke and Edwards that we must be very careful in analysing the negative findings of these experiments in the light of Mach's principle. As we have seen, Schrödinger pointed out correctly that the precession of the perihelion of the planets can be considered to be due to the anisotropy of the inertial mass. This was in a purely gravitational situation. The connection of gravitation with electromagnetism is reasonable and plausible. It is possible that gravitation and inertia come from fourth and sixth order terms in the electromagnetic potential energy ([130] and [131]). If this is the case, then the anisotropy in inertial mass may be the same as the anisotropy in electromagnetic forces, in such a way as to rule out observation of the effect in complex experiments such as these. The same can be said of nuclear forces, although their connection to gravitational and electromagnetic forces is not yet clear. What should be kept in mind is that at least in purely gravitational situations, the effect has been found, leading to the precession of the perihelion of the planets. The same can be said in electromagnetic situations, as it has been shown that self-induction of a circuit is different depending on the direction of current flow [181]. The self-induction of a circuit has been shown to be a measure of the effective inertial mass of the conduction electrons, at least according to Weber's electrodynamics.

Let us illustrate this discussion with a simple example. We assume a situation in which Eq. (8.48) is satisfied, so that the force exerted by the distant galaxies on a test body can be written as $-\Phi m_g \vec{a}_{me}$. On a frictionless table we have a body of gravitational mass m_g oscillating horizontally, connected to a spring of elastic constant k, shown in Figure 10.1.

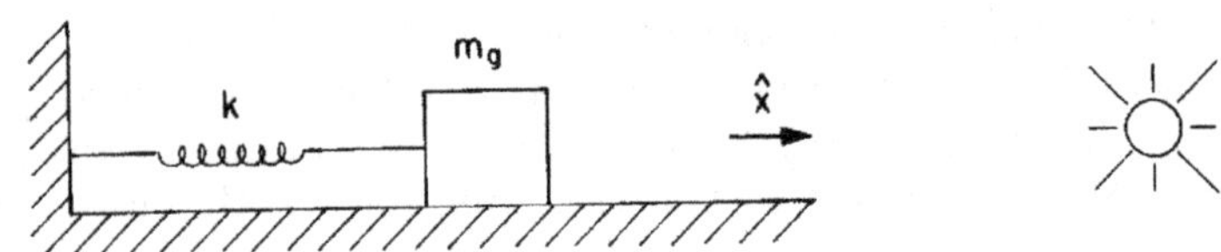

Figure 10.1: Oscillation of a body of mass m_g aligned with the sun.

The force exerted by the spring on m_g is represented by $-K\vec{r} = -Kx\hat{x}$, where $\vec{r} = x\hat{x}$ is the position vector of m_g from the point of equilibrium of the spring. The gravitational force of the earth is balanced by the normal force exerted by the table, so that we will disregard it here. Here we analyse the influence of the sun on the anisotropy of the inertial mass of m_g. In the situation of Figure 10.1 we have the sun aligned with the oscillation of the test body along the x axis. According to Weber's law the force exerted by the sun of gravitational mass M_g on m_g is given by Eq. (8.4). When the values of $\dot{r}$ and $\ddot{r}$ are used in terms of $\vec{r}_{12}$, $\vec{v}_{12}$ and $\vec{a}_{12}$ this force can be expressed as:

$$\vec{F}_{Mm} = -H_g M_g m_g \frac{\hat{r}_{12}}{r_{12}^2} \left\{ 1 + \frac{\xi}{c^2} \left[\vec{v}_{12} \cdot \vec{v}_{12} - \frac{3}{2}(\hat{r}_{12} \cdot \vec{v}_{12})^2 + \vec{r}_{12} \cdot \vec{a}_{12} \right] \right\} .$$

$$(10.12)$$

For any oscillation of the spring around the point of equilibrium we can consider $r_{12} \approx R = $ constant, where R is the earth-sun distance. As the mass of the sun is much greater than the mass m_g, we can disregard its motion relative to the frame of fixed stars, so that $\vec{r}_2 = R\hat{x}$, $\vec{v}_2 = 0$ and $\vec{a}_2 = 0$. Since the test body is oscillating along the x axis we can write: $\vec{r}_1 = x\hat{x}$, $\vec{v}_1 = \dot{x}\hat{x}$ and $\vec{a}_1 = \ddot{x}\hat{x}$ and $\vec{r}_{12} \approx -R\hat{x}$. If the velocity terms are small, $(\dot{x}/c)^2 \ll 1$, the equation of motion for m_g becomes:

$$H_g M_g m_g \frac{\hat{x}}{R^2} \left(1 - \frac{R\ddot{x}}{c^2} \right) - Kx\hat{x} - \Phi m_g \ddot{x}\hat{x} = 0 .$$

The constant force $H_g M_g m_g / R^2$ does not change the frequency of oscillation and only changes the point of equilibrium, so that we will not consider it here.

With Eqs. (8.16), (8.37), (8.45) and in these approximations, the equation of motion becomes the equation of a Newtonian harmonic oscillator given by $kx + m^*\ddot{x} = 0$, where $m^* \equiv m_g(1 + \xi GM_g/Rc^2)$ is the effective inertial mass of the test body. The solution of this equation is a sinusoidal oscillation with a frequency given by: $\omega_a = \sqrt{k/m^*}$.

Let us now consider an oscillation of the test body along the x axis, but now with the sun located along the y axis, as in Figure 10.2.

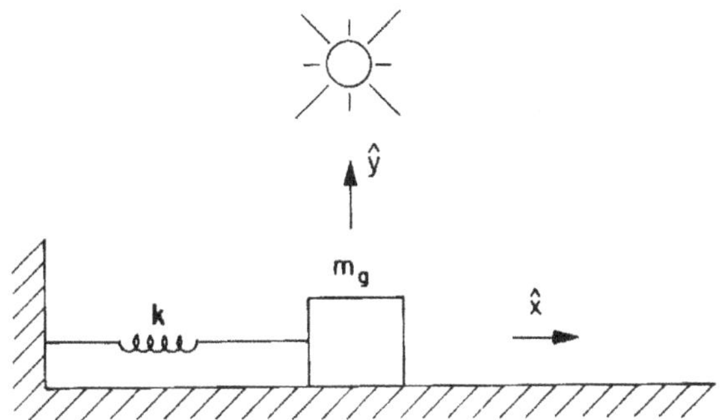

Figure 10.2: Oscillation of a body of mass m_g orthogonal to the sun.

The difference from the previous situation is that now we should approximate $\vec{r}_2 = R\hat{y}$, $\vec{r}_{12} = x\hat{x} - R\hat{y} \approx -R\hat{y}$, $\hat{r}_{12} = -\hat{y}$, where we are disregarding terms of order x/R compared with unity, namely: $x/R \ll 1$. With the previous approximation the equation of motion in the x direction becomes $kx + m_g\ddot{x} = 0$, yielding a sinusoidal solution with a frequency of oscillation given by $\omega_o = \sqrt{k/m_g}$. This example shows that the frequency of oscillation when the test body is aligned with the sun should be different from the case when the oscillation is orthogonal to the sun.

Some critical remarks are in order. This simple example illustrates very clearly the effect of a component of the force law which depends on the acceleration of the test body. The consequence is an anisotropy in the effective inertial mass, which in this case would be seen by a frequency of oscillation depending on the direction of vibration. But to arrive at this result, we had to consider several things simultaneously. First of all the analysis should be performed including the velocity terms. But they can reasonably be neglected if we observe that in Eq. (10.12) we are comparing terms of the order $\dot{x}^2$ with those of order $R\ddot{x}$. The solution of the equation is essentially $x = A\sin\omega t$, so that $\dot{x} = A\omega\cos\omega t$ and $\ddot{x} = -A\omega^2\sin\omega t$, so that $\dot{x}^2 \approx A^2\omega^2 \ll R\ddot{x} \approx RA\omega^2$, as the earth-sun distance is much larger than the amplitude of oscillation. Despite this fact it should be kept in mind that Weber's force depends not only on the acceleration between bodies, but also on their velocities, and these terms may be relevant in some experiments. Another factor was also considered simultaneously in this analysis: we supposed the elastic constant K or $k = K/\Phi$ to be the same no matter what the direction to the sun. The dimension of k is kg/s^2, so that it may happen that its value is also anisotropic. If the inertial mass of a test body is anisotropic, the same may be true of the elastic constant of a spring, as embodied in it there is also something with the dimension of mass. The same might also hold for electromagnetic and nuclear forces. If the anisotropies match those of the inertial mass, the effect would be masked. Only experiments can decide the matter here, showing whether there is an anisotropy in the frequency of oscillation. But these possibilities should be kept in mind.

Another experimental test was suggested by Eby in 1979 [194]. Essentially, he calculated the precession of a gyroscope utilizing Weber's Lagrangian energy applied to gravitation (without being aware of Weber's electrodynamics). He obtained geodetic and motional precessions which differed from those of general relativity (the Lense-Thirring effect) by factors of 2 and 3/2, respectively. His analysis should first be checked independently, and then the experiments should be performed to distinguish these models. It is interesting to quote his discussion of these predictions (our words between square brackets):

> It is conceptually satisfying that in these theories [*i.e.*, relational mechanics which he is constructing based on Weber's law] it is clear what the gyroscope is precessing with respect to, namely, the distant matter. This is not the case in metric theories of gravity [like Einstein's general theory of relativity] since there is no distant matter explicitly included in the Schwarzschild metric or its equivalent.

Another extremely important point to be tested directly is the existence of an exponential decay in gravitation. This is not necessarily connected with relational mechanics or Mach's principle, but as we have seen if we have an exponential decay in Newton's potential energy it is reasonable to suspect that an analogous term should exist which multiplies both terms of Weber's potential energy; see Eq. (8.5). Experiments to test the Seeliger-Neumann term have been performed since the last century, with some of them yielding positive results. We reviewed this subject in another study [35]. We suggest especially the repetition of Q. Majorana's many experiments on this effect ([47], [48], [49] and [50]).

Many other tests will appear in due course as more people begin working along these lines of research.

Chapter 11

History of Relational Mechanics

Now that we have presented relational mechanics and the main results we can obtain with it, let us put the main steps leading to its discovery in perspective.

As we have seen, Leibniz, Berkeley and Mach clearly visualized the main qualitative aspects of a relational mechanics. Yet none of them implemented it quantitatively. Here we present a brief history of the quantitative implementation of relational mechanics [12].

11.1 Gravitation

Although Newton had the first insights regarding gravitation in his *Anni Mirabilis* of 1666-67, the clear and complete formulation of universal gravitation seems to have come only in 1685, after a correspondence with Hooke in 1879-80 ([195, Chapter 5] and [196]). His force of gravitation appeared in print for the first time only with the publication of the *Principia* in 1687. Nowadays we write it in the form

$$\vec{F}_{21} = -Gm_{g1}m_{g2}\frac{\hat{r}_{12}}{r_{12}^2} \ .$$

Hooke and others had the idea of a gravitational force falling as the inverse square of the distance between the sun and the planets. But it is remarkable how Newton arrived at the universality of this force and the fact that it should be proportional to the product of the masses. To obtain this latter result his third law of motion, the law of action and reaction, was essential. We saw this in section 1.2, when we presented some quotations from Newton.

Newton defended the ideas of absolute space and time. Despite this fact, his force of gravitation is the first relational expression for interactions which appeared in science. It depends only on the distance between the interacting bodies and is directed along the straight line connecting them.

The introduction of the scalar potential function in gravitation is due to Lagrange (1736-1813) in 1777 and to Laplace (1749-1827) in 1782. The gravitational potential energy can be expressed as:

$$U_{12} = -G\frac{m_{g1}m_{g2}}{r_{12}} \ .$$

Once more this is completely relational. To obtain the force exerted by 2 on 1 we utilize the procedure $\vec{F}_{21} = -\nabla_1 U_{12}$.

The gravitational paradox which appears with Newton's law of gravitation in an infinite universe was discovered by H. Seeliger and C. Neumann in 1895-6. Their solution was to introduce an exponential decay in the gravitational potential of each point mass. Many others have proposed the same idea for different reasons.

11.2   Electromagnetism

Coulomb arrived at the force between point charges in 1785. There is a partial English translation of his works in Coulomb, 1935: [197] and [198]. Coulomb's force can be expressed as:

$$\vec{F}_{21} = \frac{q_1 q_2}{4\pi\varepsilon_o}\frac{\hat{r}_{12}}{r_{12}^2} \ .$$

He also arrived at an expression relating the force between two magnetic poles q_1^m and q_2^m given by:

$$\vec{F}_{21} = \frac{\mu_o}{4\pi}q_1^m q_2^m \frac{\hat{r}_{12}}{r_{12}^2} \ .$$

These two expressions are completely relational, as they have the same structure as Newton's force of gravitation.

It seems that Coulomb arrived at the force between point charges more by analogy with Newton's law of gravitation than by his doubtful measurements with the torsion balance [8]. He perfomed only three experiments of attraction and three of repulsion, but his results could not be reproduced when his experiments were repeated recently. Moreover, he never tested the proportionality of the force on $q_1 q_2$. In principle the electric forces might behave like $q_1 + q_2$, or like $(q_1 q_2)^n$ with an exponent n different from 1. Only experiments could have decided this, but he did not perform them. But in the end, his proposed force law proved to be extremely successful in explaining many phenomena. It is curious to see how he presents the fundamental law of electricity [197]: "The repulsive force between two small spheres charged with same sort of electricity is in the inverse ratio of the squares of the distances between the centers of the two spheres. Experiment: (...)" There is no mention of the proportionality in $q_1 q_2$. On the other hand, for the law of magnetic force he writes ([198]): "The magnetic fluid acts by attraction or repulsion in a ratio compounded directly of the density of the fluid and inversely of the square of the distance of its molecules. The first part of this proposition does not need to be proved; let us pass to the second. (...)" While Coulomb thinks it is not necessary to prove that the force is proportional to $q_1^m q_2^m$, Newton made a very thorough analysis before concluding that the gravitational force should be proportional to $m_1 m_2$. This at least shows a great distinction between these two scientists.

By analogy with the gravitational potential proposed by Lagrange and Laplace, Poisson introduced the scalar potential in electromagnetism in 1811-13. The energy of interaction between two point charges or between two magnetic poles is then given by

$$U_{12} = \frac{q_1 q_2}{4\pi\varepsilon_o} \frac{1}{r_{12}} \ ,$$

$$U_{12} = \frac{\mu_o}{4\pi} \frac{q_1^m q_2^m}{r_{12}} \ .$$

In 1820 Oersted discovered experimentally the deflection of a magnetized needle by a current-carrying wire. Fascinated by this fact, Ampère (1775-1836) performed a series of classical experiments and in the period between 1820-26 arrived at the following expression describing the force exerted by a current element $I_2 d\vec{l_2}$ located at $\vec{r}_2$ on another $I_1 d\vec{l_1}$ located at $\vec{r}_1$:

$$d^2\vec{F}_{21} = -\frac{\mu_o}{4\pi} I_1 I_2 \frac{\hat{r}_{12}}{r_{12}^2} \left[2(d\vec{l_1} \cdot d\vec{l_2}) - 3(\hat{r}_{12} \cdot d\vec{l_1})(\hat{r}_{12} \cdot d\vec{l_2}) \right] \ .$$

Once more this is a completely relational force. Even here the influence of Newton was very large, although this force is much more complex than the Newtonian one due to the dependence on the angles between the current elements and between the straight line connecting 1 and 2 and each one of them. To arrive at this expression Ampère assumed explicitly the proportionality of the force on $I_1 d\ell_1$ and $I_2 d\ell_2$, and also supposed that it obeys the law of action and reaction with the force along the line connecting the elements. These two facts did not emerge from the experiments. But as with Coulomb's force,

Ampère's force has been shown to be extremely successfull in explaining many phenomena of electrodynamics.

The great influence of Newton's law of gravitation upon Ampère's work can be seen in his main work summarizing his researches, *On the mathematical theory of electrodynamics phenomena, experimentally deduced* [199]. This work was published in the *Mémoires de l'Académie Royales des Sciences de Paris* for 1823. Despite this date this volume was published only in 1827. In the printed version were incorporated communications which transpired after 1823, and Ampère's paper is dated August 30th, 1826. This work has been partially translated to English by Tricker [200, pp. 155-200]. See especially the beginning of the paper and p. 172 of the English translation.

The influence of Newton's law of gravitation is clear, as Ampère assumed the force to be along the straight line connecting the elements ($\hat{r}_{12}$) and to be proportional to $I_1 d\ell_1 I_2 d\ell_2$. He then proceeded to derive from his experiments that this force between current elements should fall as r^2 and be proportional to $2(d\vec{\ell_1} \cdot d\vec{\ell_2}) - 3(\hat{r}_{12} \cdot d\vec{\ell_1})(\hat{r}_{12} \cdot d\vec{\ell_2})$.

The first to test directly the fact that the force was proportional to $I_1 I_2$ was W. Weber in 1846-48 ([201], [202] and [203]). To this end he measured directly the forces between current carrying circuits with the electrodynamometer he invented. Ampère never measured the forces directly and utilized only null methods of equilibrium which did not yield forces.

As regards the energy of interaction between two current elements, there have been many proposals. They can be summarized following Helmholtz by the expression (see Woodruff [204], Wise [205], Archibald [206], Graneau [207], Bueno and Assis [208]):

$$d^2 U_{12} = \frac{\mu_o}{4\pi} \frac{I_1 I_2}{r} \left[\frac{1+k}{2}(d\vec{l_1} \cdot d\vec{l_2}) + \frac{1-k}{2}(\hat{r} \cdot d\vec{l_1})(\hat{r} \cdot d\vec{l_2}) \right] . \qquad (11.1)$$

Here k is a dimensionless constant. Although F. Neumann worked with only closed circuits, we may say that his energy between current elements of 1845 would be given by this equation with $k = 1$. Weber's electrodynamics to be discussed next yields $k = -1$ [12, Section 4.6]. Maxwell's electrodynamics yields $k = 0$. More recently, Graneau proposed an expression like this equation with $k = 5$ ([207]). But no matter what the value of k, these are all relational energies.

Although most textbooks present Neumann's expression as representing the energy of interaction between two current elements in Maxwell's theory, this is not the case. The energy of interaction according to Maxwell should really be given by $k = 0$, and not by $k = 1$ ([204], [205], [206]). This can be seen utilizing Darwin's Lagrangian energy of 1920 which describes the interaction of two point charges q_1 and q_2 located at $\vec{r}_1$ and $\vec{r}_2$, moving with velocities $\vec{v}_1$ and

$\vec{v}_2$, respectively. It is the Lagrangian of classical electromagnetism (Maxwell-Lorentz's theory) involving relativistic corrections, time retardation and radiation effects, correct up to second order in v/c, inclusive. It is given by (Darwin [209], Jackson [210, Section 12.7, pp. 593-595], Assis [12, Section 6.8]):

$$U_{12}^D = U_{21}^D = \frac{q_1 q_2}{4\pi\varepsilon_o}\frac{1}{r_{12}}\left[1 - \frac{\vec{v}_1 \cdot \vec{v}_2 + (\vec{v}_1 \cdot \hat{r}_{12})(\vec{v}_2 \cdot \hat{r}_{12})}{2c^2}\right] .$$

Let us suppose the current elements to be composed of positive and negative charges, $dq_{1-} = -dq_{1+}$ and $dq_{2-} = -dq_{2+}$. The energy to bring the elements from an infinite distance from one another to the final separation r_{12} is given by

$$d^2 U_{12} = d^2 U_{2+,1+} + d^2 U_{2+,1-} + d^2 U_{2-,1+} + d^2 U_{2-,1-} .$$

Utilizing the fact that $I_1 \vec{dl_1} \equiv dq_{1+}\vec{v}_{1+} + dq_{1-}\vec{v}_{1-}$, $I_2 \vec{dl_2} \equiv dq_{2+}\vec{v}_{2+} + dq_{2-}\vec{v}_{2-}$, the charge neutrality of the elements and Darwin's Lagrangian yields Eq. (11.1) with $k = 0$.

Attempting to unify electrostatics with electrodynamics, so that he could derive the forces of Coulomb and Ampère from a single expression, in 1846 W. Weber proposed that the force exerted by charge q_2 on q_1 should be given by:

$$\vec{F}_{21} = \frac{q_1 q_2}{4\pi\varepsilon_o}\frac{\hat{r}_{12}}{r_{12}^2}\left(1 - \frac{\dot{r}_{12}^2}{2c^2} + \frac{r_{12}\ddot{r}_{12}}{c^2}\right) . \tag{11.2}$$

The constant $c = 3\times10^8$ m/s is the ratio of electromagnetic and electrostatic units of charge. Its value was first determined experimentally by Weber and Kohlrausch in 1856 [211], [212], [213], [204], [214], [215], [205], [216], [217, Vol. 1, pp. 144-146 and 296-297] and [218].

In 1848 he proposed an interaction energy from which this force might be derived, namely:

$$U_{12} = \frac{q_1 q_2}{4\pi\varepsilon_o}\frac{1}{r_{12}}\left(1 - \frac{\dot{r}_{12}^2}{2c^2}\right) . \tag{11.3}$$

It is important to observe that in order to arrive at Eq. (11.2) Weber began with electrostatics (Coulomb's force) and with Ampère's force between current elements. For this reason he was being influenced by Newtonian ideas, although indirectly.

Expressions (11.2) and (11.3) are once more completely relational. Despite this fact they present major differences as regards Newton's law of gravitation due to the dependence on the velocity and acceleration of the charges. This was the first time in physics that a force was proposed which depended on the velocity and acceleration between the interacting bodies. Later on many other proposals appeared in electromagnetism describing the force between point charges, such as those of Gauss (developed in 1835 but published only in 1877), Riemann (developed in 1858 but published only in 1867), Clausius in 1876 and Ritz in 1908. Further references and discussions can be found elsewhere [12, Appendix B: Alternative Formulations of Electrodynamics]. Beyond differences in form, there is a tremendous distinction between Weber's expression and all these others: Only Weber's force is completely relational, depending only on the distance, radial relative velocity and radial relative acceleration between the point charges; and thus has the same value for all observers or frames of reference. On the other hand the other expressions depend either on the velocity and acceleration of the charges relative to a preferred frame or medium like an ether, or relative to the observer.

Lorentz's force of 1895, as developed in the works of Lienard, Wiechert, Schwarzschild and Darwin, can also be written as an interaction between point charges. When this is done there also appear velocities and accelerations between the charges and the ether (as thought by Lorentz) or between the charges and inertial frames of reference (interpretation introduced by Einstein). Once more it is not the velocity and accelerations between the point charges which matter, but their motion relative to something external to them. This external frame may be the ether, the observer or a frame of reference. Only Weber's electrodynamics is completely relational. For this reason it is the only one compatible with the relational mechanics presented in this book.

In 1868 C. Neumann arrived at the Lagrangian energy describing Weber's electrodynamics, namely:

$$S_{12} = \frac{q_1 q_2}{4\pi\varepsilon_o} \frac{1}{r_{12}} \left(1 + \frac{\dot{r}_{12}^2}{2c^2} \right) . \tag{11.4}$$

The Lagrangian of a two body system might then be written as $L \equiv T - S_{12}$, where $T = m_1 v_1^2/2 + m_2 v_2^2/2$ is the kinetic energy of the system. Note the sign difference in front of $\dot{r}_{12}$ when comparing U_{12} and S_{12}. The Lagrangian energy S_{12} is also completely relational.

In 1872 Helmholtz found that the energy of a test charge q interacting with a surrounding non-conducting charged spherical shell of radius R and charge Q according to Weber's electrodynamics is given by [219]

$$U_{qQ} = \frac{qQ}{4\pi\varepsilon_o} \frac{1}{R} \left(1 - \frac{v^2}{6c^2} \right) . \tag{11.5}$$

To arrive at this expression Helmholtz supposed a stationary spherical shell interacting with an internal point charge q located anywhere inside the shell and moving with velocity $\vec{v}$ relative to the shell.

An analogous expression obtained with a Weber's law applied to gravitation is the key for the implementation of Mach's principle, as we have seen.

(Remember that Mach's ideas on mechanics had been published since 1868.) By analogy with Helmholtz's calculations, applied now to a Weberian potential energy for gravitation, this energy of interaction turns out to be exactly the kinetic energy of classical mechanics (the stars and distant galaxies would be considered as a system of spherical shells surrounding the solar system). But Helmholtz always had a negative attitude towards Weber's electrodynamics. Instead of taking this result as a hint for explaining the inertia of bodies or the origin of kinetic energy, he proposed this result as a failure of Weber's electrodynamics. Maxwell presented Helmholtz's criticisms of Weber's electrodynamics in his *Treatise* of 1873 [89, Vol. 2, Chapter 23]. He did not observe that Helmholtz's result was the key to unlock the mystery of inertia. The same can be said of all the readers of Maxwell's book at the end of last century and during this century, who had available to them not only Helmholtz's result, but Mach's books as well. We discussed this in detail in earlier work [12, Section 7.3: Charged Spherical Shell]. We can say that Helmholtz and Maxwell lost a golden opportunity to create a relational mechanics utilizing a result analogous to this one in gravitation. Fortunately, Schrödinger and others obtained similar results and were prepared to draw all the important consequences from them.

11.3 Weber's Law Applied to Gravitation

Due to the great success of Weber's electrodynamics in explaining electrostatic (through Coulomb's force) and electrodynamic phenomena (Ampère's force, Faraday's law of induction, *etc.*) some writers tried to apply an analogous expression to gravitation. The pendulum swung back: after the great influence of Newton's gravitational force on Coulomb and Ampère, it was gravitation's turn to be influenced by electromagnetism.

The idea is that the force exerted by the gravitational mass m_{g2} on m_{g1} should be given by

$$\vec{F}_{21} = -G m_{g1} m_{g2} \frac{\hat{r}_{12}}{r_{12}^2} \left(1 - \frac{\xi \dot{r}_{12}^2}{2c^2} + \frac{\xi r_{12} \ddot{r}_{12}}{c^2} \right) . \tag{11.6}$$

The energy of interaction would then be given by:

$$U_{12} = -G m_{g1} m_{g2} \frac{1}{r_{12}} \left(1 - \frac{\xi \dot{r}_{12}^2}{2c^2} \right) . \tag{11.7}$$

The gravitational Lagrangian energy is accordingly:

$$S_{12} = -G m_{g1} m_{g2} \frac{1}{r_{12}} \left(1 + \frac{\xi \dot{r}_{12}^2}{2c^2} \right) . \tag{11.8}$$

The first to propose a Weber's law fo gravitation seems to have been G. Holzmuller in 1870 [32, p. 46]. Then in 1872 Tisserand studied Weber's force applied to gravitation and its application to the precession of the perihelion of the planets. The two-body problem in Weber's electrodynamics had been solved by Seegers in 1864 [32, p. 46]. But Tisserand solved the problem iteratively, more or less as outlined in this book.

Other people also worked with Weber's law for gravitation applying it to the problem of the precession of the perihelion of the planets: Paul Gerber in 1898 and 1917, Erwin Schrödinger in 1925, Eby in 1977 and ourselves in 1989 ([220], [221], [125] with English translation in [83], [222] and [132]). Curiously, none of them were aware of Weber's electrodynamics, with the exception of our work. Each one of them arrived at Eqs. (11.7) or (11.8) on his own. Gerber was working with ideas of retarded time and worked in the Lagrangian formulation. Schrödinger was trying to implement Mach's principle with a relational theory. Eby was following the works of Barbour on Mach's principle and also worked with the Lagrangian formulation.

Poincaré discussed Tisserand's work on Weber's law applied to gravitation in 1906-7 [223, pp. 125 and 201-203]. Gerber's works were criticized by Seeliger, who was aware of Weber's electrodynamics [224].

For references to other writers who have applied Weber's law to gravitation in the second half of this century, see Assis [12, Section 7.5].

It should be emphasized that Weber himself considered the application of his force law to gravitation. Working in collaboration with F. Zollner in the 1870's and 1880's he applied the ideas of Young and Mossotti of deriving gravitation from electromagnetism. But instead of working with Coulomb's force they employed Weber's own force between point charges, so that the final result was a Weber's law applied to gravitation (Young and Mossotti had arrived at a similar to Newton's law of universal gravitation). For references see Woodruff [215] and Wise [205]. We are not sure if they published works prior to those of Holzmuller and Tisserand in 1870 and 1872.

With the exception of Schrödinger, Eby and ourselves, the other authors who applied Weber's law to gravitation we have quoted here were not concerned with Mach's principle.

11.4   Relational Mechanics

Mach suggested that the inertia of a body should be connected with distant matter and specially with the fixed stars (in his time the external galaxies were not yet known). He did not discuss or emphasize the proportionality between inertial and gravitational masses. He did not say that inertia should be connected with a *gravitational* interaction with distant masses. He did not propose any specific force law to implement his ideas quantitatively (for instance, showing that a spinning set of stars generates centrifugal forces). However, his book *The Science of Mechanics* was extremely influential as regards physics, much more than Leibniz's or Berkeley's writings. It was published in 1883, and from that time onwards people began trying to implement his intuitive ideas, which were very appealing.

The first to propose a Weber's law for gravitation in order to implement Mach's principle seems to have been I. Friedlaender in 1896. This suggestion appeared in a footnote on page 17 of the book by the Frielander brothers, in which each part was written by a brother [78]. A partial English translation of this book can be found in Friedlaender and Friedlaender, 1995 [79]. Immanuel Friedlaender begins by speaking about the centrifugal force (tendency to depart from the axis of rotation) which appears when we spin an object relative to the earth. He says that it should be possible to reverse this. The centrifugal force should appear when we rotate the earth and the distant universe in the opposite sense relative to the test body. He believes Newtonian mechanics is incomplete as it does not supply this equivalence. Then comes the part which concerns us here [79, see especially pp. 310-311]: "(...) it seems to me that the correct form of the law of inertia will only then have been found when *relative inertia* as an effect of masses on each other and *gravitation*, which is also an effect of masses on each other, have been derived on the basis of a *unified law*.[1] The challenge to theoreticians and calculators to attempt this will only be crowned with success when the invertibility of centrifugal force has been successfully demonstrated. Berlin, New Year 1896."

This was only a suggestion and they did not develop it further. Despite this fact it was important in at least two respects: They were the first to suggest in print that inertia is due to a *gravitational* interaction. Moreover, they proposed Weber's law as the kind of interaction to work with. The inversion of the centrifugal force (the dynamical equivalence for kinematically equivalent situations), has been completely implemented in relational mechanics, as we have seen in this book.

In 1900 Höfler also suggested an application of Weber's law for gravitation in order to implement Mach's principle, [68, pp. 21 and 41]. Once more, this suggestion was not developed.

In 1904 W. Hofmann proposed to replace the kinetic energy $m_i v^2/2$ by a two body interaction like $L = kMmf(r)v^2$, where k is a constant, $f(r)$ some function of the distance between the bodies of masses M and m, and

---

[1] In this connection it is greatly to be desired that the question of whether Weber's law is to be applied to gravitation and also the question of the propagation velocity of gravitation should be resolved. For the second issue, one could use an instrument that makes it possible to measure statically the diurnal variations of the earth's gravity as a function of the position of the heavenly bodies.

v is the relative speed between M and m. His work is discussed in Norton [68]. The usual result $mv^2/2$ would be recovered after integrating L over all masses in the universe. Hofmann did not complete the implementation of his qualitative idea. His work is important because he is considering an interaction of Weber's type (see Eq. (11.7)) to arrive at the kinetic energy, although he did not specify the function $f(r)$. However, he did not seem to be aware of Weber's electrodynamics.

Although Einstein was greatly influenced by Mach's book on mechanics, he did not try to employ a relational expression for the energy or force between masses. He never mentioned Weber's force or potential energy. All those who were influenced by Einstein's line of reasoning remained very far from relational mechanics. For this reason we do not consider them here.

After the Friedlaenders, Höfler and Hofmann, another important person who attempted to implement Mach's principle utilizing relational quantities was Reissner. Without being aware of Weber's work he arrived independently at a potential energy very similar to Weber's potential applied to gravitation [225] with English translation in [81], and [226] with partial English translation in [82]. In the article of 1914 he works with a classical gravitational potential energy plus a term of the type $m_1 m_2 f(r)\dot{r}^2$, particularized for $f(r) = $ constant. In 1915 he substitutes this term for $m_1 m_2 \dot{r}^2/r$. Unfortunately, from 1916 onwards he began to develop Einstein's ideas on general relativity and no longer worked with relational quantities [227] and [68, p. 33].

Erwin Schrödinger (1887-1961) wrote a very important paper in 1925 where he arrived at the main results of relational mechanics [125] with English translation in [83].

In this paper Schrödinger says that he wishes to implement Mach's ideas. He mentions the fact that Einstein's general theory of relativity does not implement these ideas and for this reason he tries a different approach. Taking the form of the kinetic energy $mv^2/2$ as a guiding idea he proposes heuristically a modified form of the Newtonian potential energy, namely:

$$U = -Gm_1 m_2(1 - \gamma \dot{r}^2)/r \ . \tag{11.9}$$

To arrive at this expression he explicitly emphasized the aspect that any interaction energy should depend only on the distance and relative velocities between the particles in order to follow Mach's approach. Absolute velocities should not appear, only relational quantities. Curiously enough, he never mentions Weber's name or Weber's law, although he was a German speaker. He integrates this energy of interaction for a spherical shell of mass M and radius R interacting with an internal point mass m located near its center and moving relative to it with velocity v and obtains the approximate result

$$U = -G\frac{mM}{R}\left(1 - \frac{\gamma v^2}{3}\right) .$$

He did not know this, but his approximate result was exact and valid anywhere inside the shell and not only near its center, as had been known since Helmholtz in 1872 (working with charges instead of masses, but the consequence is the same). Schrödinger identifies this result with the kinetic energy of the test body and implicitly arrives at the main results of relational mechanics. He then considers a "two-body problem" (the sun, the planet and the distant masses) and arrives at the precession of the perihelion of the planets. As we have seen, others had arrived at this result before him, but he does not quote anyone. To get the Einstenian relativistic result, which was known to agree with the observed values, Schrödinger obtains $\gamma = 3/c^2$. Then he integrates the result of the spherical shell for the whole world up to a radius R_o supposing a constant mass density ρ_o, and obtains relation (8.37), namely: $G = c^2/4\pi\rho_o R_o^2$. He observes that taking R_o and ρ_o as the radius and density of our galaxy we would obtain a value of G 10^{11} times smaller than what is observed. His conclusion is then that the inertia of bodies in the solar system is due mainly to matter farther away from our galaxy. For Eq. (8.37) to be valid, with $G = 6.67 \times 10^{-11}$ Nm^2/kg^2, R_o needed to be much greater than all other astronomical distances known at his time. It is curious to observe that the existence of external galaxies had just been confirmed by E. Hubble in 1924. Until then, many thought the whole universe was only our own galaxy. Hubble's law of redshifts appeared only in 1929. This relation between G, R_o, c and ρ_o was rediscovered later by Sciama in 1953, by Brown in 1955, by Edwards in 1974, by Eby in 1977 and by ourselves in 1989 [70], [71], [193], [222], [132].

Schrödinger then goes a step further. He takes the classical kinetic energy as an approximation for small velocities and assumes the relativistic kinetic energy $mc^2(1/\sqrt{1 - v^2/c^2} - 1)$ as an empirical relation valid for low and high velocities. To derive this energy he modifies Weber's potential energy by the expression

$$U = -G\frac{m_{g1}m_{g2}}{r}\left(3 - \frac{2}{(1 - \dot{r}^2/c^2)^{3/2}}\right) .$$

This reduces to Weber's expression up to second order in $\dot{r}/c$. After integrating this expression for the distant masses analogously to the prior procedure, Schrödinger obtains an expression like the relativistic kinetic energy. He also observes that this energy can be derived from a Lagrangian energy L given by

$$L = G\frac{m_{g1}m_{g2}}{r}\left(\frac{2}{\sqrt{1 - \dot{r}^2/c^2}} - 4\sqrt{1 - \dot{r}^2/c^2} + 3\right) .$$

After the usual integration, this yields a Lagrangian analogous to the relativistic Lagrangian for a point particle, namely: $L = -mc^2\sqrt{1 - v^2/c^2}$. Once more the mass which appears in Schrödinger's derivation is the gravitational mass. Moreover, the velocity of the test charge is relative to the frame of distant matter (frame in which the distant masses are at rest).

Recently Wesley published a similar work, without being aware of Schrödinger's paper [184].

To our knowledge, he did no more work along these lines after this paper, nor had he published anything previously on this subject. This was one of his last paper before the famous works on quantum mechanics, where he developed Schrödinger's equation and the wave approach to quantum mechanics. The enormous success of these papers may explain why he did not return to his work on Mach's principle.

Another reason may have to do with Reissner. In his article of 1925 Schrödinger claims that he arrived at Eq. (11.9) "heuristically." He does not mention Weber, Tisserand, Reissner, nor any other author. Now if indeed he did achieve this heuristically, he should have arrived at this expression all by himself. Let us quote here the relevant passage [83] (our emphasis):

> One must therefore see if it is possible in the case of the kinetic energy, just as hitherto for the potential energy, to assign it, not to mass points individually, but instead also represent it as an energy of interaction of any two mass points and let it depend only on the separation and the rate of change of the separation of the two points. In order to select an expression from the copious possibilities, we use *heuristically* the following analogy requirements:
>
> 1. The kinetic energy as an interaction energy shall depend on the masses and the separation of the two points in the same manner as does the Newtonian potential.
>
> 2. It shall be proportional to the square of the rate of change of the separation.
>
> For the total interaction energy of two mass points with the masses μ and μ' with separation r we then obtain the expression
>
> $$W = \gamma\frac{\mu\mu'\dot{r}^2}{r} - \frac{\mu\mu'}{r}\ .$$
>
> The masses are here measured in a unit such that the gravitational constant has the value 1. The constant γ, which for the moment is undetermined, has the dimensions of a reciprocal velocity. Since it should be universal, one will expect that, apart from a numerical

factor, this will be the velocity of light, or that γ will be reduced to a numerical factor when the light second is chosen as the unit of time. We shall have cause later to set this numerical factor equal to 3.

As a matter of fact, this is not the whole history of how Schrödinger arrived at this relation. The collected works of Schrödinger have been published recently. At the end of the reprint of this article there is a typewritten note, signed by Schrödinger, where he expresses apologies to Reissner for plagiarizing his ideas, unconsciously [228, p. 192]. He says that he knew his first paper of 1914 but is not certain as regards the second one of 1915. Perhaps the fact that he utilized Reissner's ideas, without quoting him, and the constraint he may have felt to admit this publicly influenced him not to deal with this subject further (others may have perceived the similarities between their works).

In any event it is a great irony that Weber's law for electromagnetism had been published some 70 years before Reissner (80 years before Schrödinger). An application of Weber's law to gravitation dates back at least to the 1870's, some 40 years before Reissner. Weber published in German, like Reissner and Schrödinger; his work was discussed by Maxwell and many others. It is amazing that Reissner and Schrödinger did not know about his work and that even after their publications in 1915 and 1925 no one called their attention to Weber's works.

There is a possible third reason why Schrödinger stopped working with a Weber's law in order to implement Mach's principle: he turned to Einstein's general theory of relativity, as had happened with Reissner. Schrödinger, for instance, worked later with a unified theory based on Einstein's works [229]. He even published a book on the expanding universe, based on Einstein's general theory of relativity [230].

This paper of 1925 was not followed or developed by other workers either. It was forgotten for the next 60 years, until it was reprinted in 1984. We found only one reference to it in another place, in a paper of 1987 [231, see especially p. 1157]. Another quotation can be found in Mehra's book [231, pp. 372-373 and 459]. Only in 1993 did it begin to be rediscovered by other people. Julian Barbour told us about this paper in July 1993, and he himself was informed about this paper by Domenico Giulini, who found it in Schrödinger's collected works (private communication by Julian Barbour and [69, p. 5]). This article was then discussed at a conference on Mach's principle which happened in Tübingen, Germany, in 1993 [69]. In the Proceedings of this conference there is a complete English translation of the paper. We published a Portuguese translation in 1994 [232]. Further applications of this approach can be found elsewhere [12, Section 7.7] and [14].

Although many important results of relational mechanics are contained in

Schrödinger's paper, he did not show that a rotating spherical shell generates inside itself centrifugal and Coriolis forces. He also did not discuss in greater detail the proportionality between inertial and gravitational masses. Moreover, as he worked only with energies, he did not derive a law analogous to the Newtonian $m\vec{a}$, nor did he discuss how to do this. He also did not know that the energy of interaction of a test particle inside a spherical shell was valid anywhere inside the shell, and not only close to its center.

After Schrödinger, we are not aware of any relational theory that seeks to implement Mach's principle for the next fifty years. Although there have been alternatives to general relativity, they were usually modelled on Einstein's approach and so maintained most non-Machian aspects of his theories of relativity (absolute quantities, inertia due to space, frame dependent forces, *etc.*) For this reason we will not consider them here.

An exception which must be mentioned is the work of Burniston Brown [71] and [233]. He did not follow general relativity but an analogy with the electromagnetic forces. Unfortunately the force expression he employed for gravitation was not exactly relational, as is the case with Weber's law. Despite this fact he arrived at several Machian consequences with his model.

In 1974 Edwards was led by analogies between electromagnetism and gravitation to work with relational quantities such as $\dot{r}$, *etc.*: [193]. He was not aware of Schrödinger's approach. He mentions that his "approach employs some of the basic ideas of Weber's and Riemann's electromagnetic theories." He draws attention to an interesting possible explanation of the origin of binding forces within fundamental particles and nuclei utilizing the fact that Weber's force applied to electromagnetism depends on the acceleration between the charges. This means that the effective inertial mass of a charged particle depends on its electrostatic potential energy, so that this effective inertial mass can become negative under certain conditions. As a consequence of this, negative charges might attract one another when these conditions are satisfied. As we have seen, Helmholtz had arrived at these ideas of an effective inertial mass depending on the electrostatic potential energy 100 years before [12]. Edwards published nothing else along these lines of implementing Mach's principle from a Weber's force applied to gravitation.

At the same time Barbour, and later Barbour and Bertotti, worked with relational quantities, intrinsic derivatives and with the relative configuration space of the universe [234], [235] and [236]. They now follow Einstein's approach closely.

Eby followed Barbour's ideas and worked with a Lagrangian energy like Eq. (11.4) applied to gravitation [222]. He calculated the precession of the perihelion of the planets with this Lagrangian and also implemented Mach's principle. Once more, he was not aware of Weber's electrodynamics or of Schrödinger's paper. In a following paper, Eby considered the precession of a gyroscope with his model, and showed that there are different predictions between relational mechanics and Einstein's general theory of relativity in the geodetic and motional precessions [194]. Recently he published another paper on Mach's principle [182].

Our own work on relational mechanics and Weber's law applied to electromagnetism and gravitation began in 1988 and is being published in several places. We have five books on these subjects ([11], [12], [13], [237] and [2]).

To our knowledge we were the first to obtain Eq. (8.12) [132]. In other words, we were the first to implement quantitatively Mach's idea that spinning the distant universe yields real centrifugal and Coriolis forces. It seems to us that no one had derived this key result before. We were also the first to derive Eq. (8.11) with $\vec{\omega} \neq 0$. Helmholtz and Schrödinger obtained it before us when $\vec{\omega} = 0$. We were also the first to derive Eqs. (8.13) and (8.14) [132], [35] and [12, Chapter 7]. We were also the first to introduce the exponential decay in Weber's potential energy, Eqs. (8.5) and (8.6) [35].

As regards the principle of dynamical equilibrium (the third postulate of relational mechanics), Sciama seems to have been the first to state a particular form of this assumption [70]. Let us quote his main postulate: "(...) in the rest-frame of any body the total gravitational field at the body arising from all the other matter in the universe is zero." The first limitation of his formulation was that he assumed it to be valid only for gravitational interactions, while we have applied it to all kinds of interaction. But much more serious than that was the fact that he restricted the validity of his postulate only to the rest frame of the test body which experiences the interaction, while we have supposed it to be valid in all frames of reference. The reason for his limited supposition is very simple. He utilized as his force law an expression similar to Lorentz's force law applied to gravitation, which is certainly not relational. Moreover, as is well known, Lorentz's force depends on the position and velocity of the test body, but not on its acceleration. When the test body was accelerated relative to the distant galaxies, Sciama was able to show, in the frame of the test body (frame always fixed with it) that the distant galaxies would exert a force on the test body of gravitational mass m_g given by $m_g \vec{a}_{Um}$, where $\vec{a}_{Um}$ is the acceleration of the set of distant galaxies relative to the test body. But in the frame of the distant galaxies there is no force exerted by them on the accelerated test body in Sciama's calculation! If you are in the universal frame of reference (fixed relative to the set of distant galaxies) and calculate the gravitational force exerted by these galaxies on a test body which is accelerated relative to them, this yields a zero value with Sciama's expression for the gravitational force (analogous to Lorentz's force), no matter what the acceleration of the test body relative to the distant galaxies. This is due to the fact that Lorentz's force is not relational, yielding different results in different frames of reference, and also because it depends on the acceleration of the source body but not on the acceleration of the test body [17], [18] and [12, Sections 6.4 and 7.3]. This means that he could not implement Mach's principle in its full generality. First of all, he did not work with relational quantities. Nor could he derive Newton's second law of motion in the frame of distant matter, where it is known to be valid. The first presentation of the principle of dynamical equilibrium in its full generality, deriving all important consequences from it, was given in our paper of 1989 [132].

Chapter 12

Conclusion

We believe strongly in the relational mechanics as presented in this book. We have written it to show this formulation in its full generality, so that others can see the power of this approach.

Graneau is one of those who grasped all aspects of relational mechanics [238], [239], [240], [241], [242] and [243, Chapter 3, The Riddle of Inertia]. Other people we can mention are Wesley [184] and [93, Chapter 6], Zylbersztajn [244] and Phipps [92].

We believe that the three postulates of relational mechanics will not need to be modified. On the other hand, experimental findings may modify Weber's law applied to gravitation and electromagnetism. For instance, it may be found necessary to introduce terms which depend on $d^3 r_{12}/dt^3$, $d^4 r_{12}/dt^4$, etc. Higher order powers of the derivatives may also appear, like: $\dot{r}$, $\dot{r}^3$, $\dot{r}^4$, ..., $\ddot{r}^2$, $\ddot{r}^3$, ... $(d^3 r/dt^3)^m$, etc. A possible exponential decay in gravitation (and maybe in electromagnetism) needs to be confirmed experimentally.

But the main lines of approaching future problems have already been laid down: no absolute space and time; only relational quantities should be involved; all forces should come from interactions between material bodies; for point particles the force should be directed along the line joining them and should obey the principle of action and reaction; *etc.*

Newton created the best possible mechanics of his time. He understood clearly the difference between inertia and weight (or between gravitational and inertial mass). He knew Galileo's result on the equality of the acceleration of freely falling bodies and performed a very accurate experiment with pendulums which showed that the inertia of a body was proportional to its weight to one part in a thousand. Although he could not explain this proportionality, he was a giant to see the importance of this fact and to perform such a precise experiment. He introduced the universal law of gravitation, which falls as the inverse square of the distance. Moreover, he proved two key theorems: a spherical shell attracts an external material particle as if it were concentrated at its center, and exerts no net force on any internal body. Both theorems are valid no matter what the motion of the test body or of the shell. He performed the bucket experiment and observed that the concavity of the water was not due to its rotation relative to the bucket. Due to his two theorems stated above, he believed the concavity of the water could not be due to its rotation relative to the earth or to the fixed stars. He had no other alternative to explain this experiment except to say that it proved the existence of absolute space which had no connections with any matter.

It was only 160 years later that Wilhelm Weber proposed an electromagnetic force depending on the distance between the point charges, on their relative radial velocity and on their relative radial acceleration. He also proposed a potential energy depending on the distance and radial relative velocity between the charges. These were the first force and energy in physics depending on velocity and acceleration between the interacting bodies. Weber's formulation is the only theory of electrodynamics ever proposed depending on relational magnitudes between the interacting charges. For this reason Weber's force and energy always have the same value in all frames of reference, even for non-inertial frames (in the Newtonian sense of this word). Weber's force complies with the principle of action and reaction. Moreover, it is directed along the straight line connecting the charges. It follows the principles of conservation of linear momentum, of angular momentum and of energy. When there is no motion between the charges, we derive from it Coulomb's force and Gauss's law of electrostatics. With Weber's force we also derive Ampère's force between current elements. From this last expression we derive the law of non-existence of magnetic monopoles and the magnetic circuital law. With his expression Weber also derived Faraday's law of induction. Weber and Kirchhoff derived, before Maxwell, a wave equation describing the propagation of electromagnetic perturbations along wires at light velocity. They worked independently from one another, but both of them based on Weber's electrodynamics. Weber was also the first to measure the electromagnetic quantity $1/\sqrt{\mu_o \varepsilon_o}$, finding the same value as light velocity in vacuum. This was one of the first quantitative indications showing a connection between optics and electromagnetism.

With a Weber's potential energy for gravitation and applying it for the interaction of a test particle and the distant universe we obtain an energy analogous to the classical kinetic energy. In this way this last expression can be seen as an interaction energy, like all other energies (elastic, electromagnetic, *etc.*) A Weber's force applied to gravitation shows that the distant universe exerts a gravitational force on any body accelerated relative to it. This force is proportional to the gravitational mass of the test body and to its acceleration relative to the distant universe. This result, together with the principle of dynamical equilibrium, yields equations of motion similar to Newton's first and second laws. Finally it explains the proportionality of inertia and weight. We could also derive the fact that the best inertial frame available to us is the distant galaxies. We have been able to explain the coincidence of Newtonian mechanics that the universe as a whole does not rotate relative to absolute space or to any inertial frame of reference. In other words, we have explained why the kinematical rotation of the earth is identical to its dynamical rotation. We have derived a relation connecting microscopic quantities (G) with macroscopic quantities (H_o and ρ_o). This relation had been known for a long time, with no convincing explanation for its origin. We have found a complete equivalence between the Ptolemaic and Copernican world systems. It is then equally valid to say that the earth moves relative to the distant universe, or that it is at rest and that it is the distant universe which moves relative to the earth. We have derived the fact that all inertial forces of Newtonian mechanics, like the centrifugal or Coriolis forces, are real forces acting between the test body and the distant universe. These forces have a gravitational origin and appear when there is a relative rotation between the body and the universe. This also explains the concavity in Newton's bucket as due to a relative rotation between the water and the distant universe, as had been suggested by Mach.

We have reached a clear and satisfactory understanding of the key facts of classical mechanics. From now on the best alternative is to follow the new path this relational approach opens up. It is the path to a new world!

Bibliography

[1] T. S. Kuhn. *The Structure of Scientific Revolutions.* The University of Chicago Press, Chicago, 1962.

[2] A. K. T. Assis. *Mecânica Relacional.* Editora do CLE da UNICAMP, Campinas, 1998. ISBN: 85-86497-01-0.

[3] I. Newton. *Mathematical Principles of Natural Philosophy.* University of California Press, Berkeley, 1934. Cajori edition.

[4] J. B. Barbour. *Absolute or Relative Motion? - A study from a Machian point of view of the discovery and the structure of dynamical theories,* volume 1: *The Discovery of Dynamics.* Cambridge University Press, Cambridge, 1989.

[5] H. Cavendish. Experiments to determine the density of the earth. *Philosophical Transactions,* 88:469–526, 1798. Reprinted in: *The Scientific Papers of the Honorable Henry Cavendish, F. R. S.,* vol. II, E. Thorpe *et al.* (editors), (Cambridge University Press, Cambridge, 1928), pp. 249-286; also reprinted with explanatory notes by A. S. Mackenzie in: *Gravitation, Heat and X-Rays,* (Arno Press, New York, 1981), pp. 57-105.

[6] B. E. Clotferter. The Cavendish experiment as Cavendish knew it. *American Journal of Physics,* 55:210–213, 1987.

[7] I. Newton. *Opticks.* Dover, New York, 1979.

[8] P. Heering. On Coulomb's inverse square law. *American Journal of Physics,* 60:988–994, 1992.

[9] A. K. T. Assis. Weber's law and mass variation. *Physics Letters A,* 136:277–280, 1989.

[10] A. K. T. Assis and J. J. Caluzi. A limitation of Weber's law. *Physics Letters A,* 160:25–30, 1991.

[11] A. K. T. Assis. *Curso de Eletrodinâmica de Weber*. Setor de Publicações do Instituto de Física da Universidade Estadual de Campinas - Unicamp, Campinas, 1992. Notas de Física IFGW Número 5.

[12] A. K. T. Assis. *Weber's Electrodynamics*. Kluwer Academic Publishers, Dordrecht, 1994. ISBN: 0-7923-3137-0.

[13] A. K. T. Assis. *Eletrodinâmica de Weber - Teoria, Aplicações e Exercícios*. Editora da Universidade Estadual de Campinas - UNICAMP, Campinas, 1995. ISBN: 85-268-0358-1.

[14] J. J. Caluzi and A. K. T. Assis. Schrödinger's potential energy and Weber's electrodynamics. *General Relativity and Gravitation*, 27:429–437, 1995.

[15] J. J. Caluzi and A. K. T. Assis. An analysis of Phipps's potential energy. *Journal of the Franklin Institute B*, 332:747–753, 1995.

[16] P. W. Worden and C. W. F. Everitt. Resource letter GI-1: gravity and inertia. *American Journal of Physics*, 50:494–500, 1982.

[17] A. K. T. Assis. Centrifugal electrical force. *Communications in Theoretical Physics*, 18:475–478, 1992.

[18] A. K. T. Assis. Changing the inertial mass of a charged particle. *Journal of the Physical Society of Japan*, 62:1418–1422, 1993.

[19] A. K. T. Assis and D. S. Thober. Unipolar induction and Weber's electrodynamics. In M. Barone and F. Selleri, editors, *Frontiers of Fundamental Physics*, pages 409–414, New York, 1994. Plenum Press.

[20] Archimedes. *The Works of Archimedes*. Dover, New York, 1912. Edited in modern notation with introductory chapters by T. L. Heath, with a supplement *The Method of Archimedes*.

[21] Archimedes. *The Works of Archimedes*. Great Books of the Western World. Encyclopaedia Britannica, Chicago, 1952. Translated by T. L. Heath.

[22] T. S. Kuhn. *The Copernican Revolution*. Harvard University Press, Cambridge, 1957.

[23] P. Caldirola and E. Recami. The concept of time in physics. *Epistemologia*, 1:263–304, 1978.

[24] G. J. Whitrow. *Time in History*. Oxford University Press, Oxford, 1989.

[25] F. A. Mella. *La Misura del Tempo nel Tempo*. Hoepli, Milano, 1990.

[26] A. A. Penzias and R. W. Wilson. A measurement of excess antenna temperature at 4080 Mc/s. *Astrophysical Journal*, 142:419–421, 1965.

[27] K. R. Symon. *Mechanics*. Addison-Wesley, Reading, third edition, 1971.

[28] L. Foucault. Démonstration physique du mouvement de rotation de la terre au moyen du pendule. *Comptes Rendues de l'Academie des Sciences de Paris*, Feb. 03:135–138, 1851.

[29] L. Foucault. Physical demonstration of the rotation of the earth by means of the pendulum. *Journal of the Franklin Institute*, 21:350–353, 1851.

[30] R. Crane. The Foucault pendulum as a murder weapon and a physicist's delight. *The Physics Teacher*, 28:264–269, 1990.

[31] M. Born. *Einstein's Theory of Relativity*. Dover, New York, Revised edition, 1965.

[32] J. D. North. *The Measure of the Universe — A History of Modern Cosmology*. Clarendon Press, Oxford, 1965.

[33] M. Jammer. *Concepts of Space - The History of Theories of Space in Physics*. Harvard University Press, Cambridge, 2nd edition, 1969.

[34] S. L. Jaki. *Cosmos in Transition — Studies in the History of Cosmology*. Pachart Publishing House, Tucson, 1990.

[35] A. K. T. Assis. On the absorption of gravity. *Apeiron*, 13:3–11, 1992.

[36] A. K. T. Assis. A steady-state cosmology. In H. C. Arp, C. R. Keys, and K. Rudnicki, editors, *Progress in New Cosmologies: Beyond the Big Bang*, pages 153–167, New York, 1993. Plenum Press.

[37] E. Harrison. Newton and the infinite universe. *Physics Today*, 39:24–32, 1986.

[38] I. B. Cohen (ed.). *Isaac Newton's Papers & Letters on Natural Philosophy*. Harvard University Press, Cambridge, second edition, 1978.

[39] E. Mach. *The Science of Mechanics — A Critical and Historical Account of Its Development*. Open Court, La Salle, 1960.

[40] V. C. Rubin, W. K. Ford, Jr., and N. Thonnard. Rotational properties of 23 Sb galaxies. *Astrophysical Journal*, 261:439–456, 1982.

[41] V. C. Rubin. Dark matter in spiral galaxies. *Scientific American*, 248:88–101, 1983.

[42] R. H. Sanders. Anti-gravity and galaxy rotation curves. *Astronomy and Astrophysics*, 136:L21–L23, 1984.

[43] R. H. Sanders. Finite length-scale anti-gravity and observations of mass discrepancies in galaxies. *Astronomy and Astrophysics*, 154:135–144, 1986.

[44] R. H. Sanders. Mass discrepancies in galaxies: dark matter and alternatives. *Astronomy and Astrophysics Review*, 2:1–28, 1990.

[45] D. S. L. Soares. An alternative view of flat rotation curves in spiral galaxies. *Revista Mexicana de Astronomia e Astrofisica*, 24:3–7, 1992.

[46] D. S. L. Soares. An alternative view of flat rotation curves. II. The observations. *Revista Mexicana de Astronomia e Astrofisica*, 28:35–42, 1994.

[47] Q. Majorana. On gravitation — Theoretical and experimental researches. *Philosophical Magazine*, 39:488–504, 1920.

[48] Q. Majorana. Quelques recherches sur l'absorption de la gravitation par la matière. *Journal de Physique et de Radium*, 1:314–324, 1930.

[49] Q. Majorana. On gravitation — Theoretical and experimental researches. In V. DeSabbata and V. N. Melnikov, editors, *Gravitational Measurements, Fundamental Metrology and Constants*, pages 523–539, Dordrecht, 1988. Kluwer.

[50] Q. Majorana. Quelques recherches sur l'absorption de la gravitation par la matière. In V. DeSabbata and V. N. Melnikov, editors, *Gravitational Measurements, Fundamental Metrology and Constants*, pages 508–522, Dordrecht, 1988. Kluwer.

[51] G. Dragoni. On Quirino Majorana's papers regarding gravitational absorption. In V. DeSabbata and V. N. Melnikov, editors, *Gravitational Measurements, Fundamental Metrology and Constants*, pages 501–507, Dordrecht, 1988. Kluwer.

[52] G. T. Gillies. Resource letter MNG-1: Measurements of Newtonian gravitation. *American Journal of Physics*, 58:525–534, 1990.

[53] R. de A. Martins. Huygens's reaction to Newton's gravitational theory. In J. V. Field and A. J. L. James, editors, *Renaissance and Revolution*, pages 203–213. Cambridge University Press, Cambridge, 1993.

[54] H. G. Alexander (ed.). *The Leibniz-Clarke Correspondence*. Manchester University Press, Manchester, 1984.

[55] G. W. Leibniz. *Philosophical Essays*. Hackett Publishing Company, Indianapolis, 1989. Edited and translated by R. Ariew and D. Garber.

[56] H. Erlichson. The Leibniz-Clarke controversy: absolute versus relative space and time. *American Journal of Physics*, 35:89–98, 1967.

[57] G. Berkeley. De Motu - Of Motion, or the principle and nature of motion and the cause of the communication of motions. In M. R. Ayers, editor, *George Berkeley's Philosophical Works*, pages 211–227, London, 1992. Everyman's Library.

[58] G. J. Whitrow. Berkeley's philosophy of motion. *British Journal for the Philosophy of Science*, 4:37–45, 1953.

[59] K. R. Popper. A note on Berkeley as precursor of Mach. *British Journal for the Philosophy of Science*, 4:26–36, 1953.

[60] K. P. Winkler. Berkeley, Newton and the stars. *Studies in History and Philosophy of Science*, 17:23–42, 1986.

[61] G. Berkeley. A Treatise concerning the Principles of Human Knowledge. In M. R. Ayers, editor, *Geroge Berkeley's Philosophical Works*, pages 61–127, London, 1992. Everyman's Library.

[62] M. Ghins. *A Inércia e o Espaço-Tempo Absoluto*. Coleção CLE. Centro de Lógica, Epistemologia e História da Ciência da UNICAMP, Campinas, 1991.

[63] J. T. Blackmore. *Ernst Mach - His Life, Work, and Influence*. University of California Press, Berkeley, 1972.

[64] L. I. Schiff. Observational basis of Mach's principle. *Reviews of Modern Physics*, 36:510–511, 1964.

[65] E. Mach. History and Root of the Principle of the Conservation of Energy. In I. B. Cohen, editor, *The Conservation of Energy and the Principle of Least Action*, New York, 1981. Arno Press. Reprint of 1911 English translation.

[66] W. Yourgrau and A. van der Merwe. Did Ernst Mach 'miss the target'? *Synthese*, 18:234–250, 1968.

[67] A. K. T. Assis. Compliance of a Weber's force law for gravitation with Mach's principle. In P. N. Kropotkin *et al.*, editor, *Space and Time Problems in Modern Natural Science, Part II*, pages 263–270, St.-Petersburg, 1993. Tomsk Scientific Center of the Russian Academy of Sciences. Series: "The Universe Investigation Problems," Issue 16.

[68] J. D. Norton. Mach's principle before Einstein. In J. B. Barbour and H. Pfister, editors, *Mach's Principle - From Newton's Bucket to Quantum Gravity*, pages 9–57, Boston, 1995. Birkhäuser.

[69] J. B. Barbour and H. Pfister (editors). *Mach's Principle: From Newton's Bucket to Quantum Gravity*. Birkhäuser, Boston, 1995.

[70] D. W. Sciama. On the origin of inertia. *Monthly Notices of the Royal Astronomical Society*, 113:34–42, 1953.

[71] G. B. Brown. A theory of action-at-a-distance. *Proceedings of the Physical Society B*, 68:672–678, 1955.

[72] F. A. Kaempffer. On possible realizations of Mach's program. *Canadian Journal of Physics*, 36:151–159, 1958.

[73] P. Moon and D. E. Spencer. Mach's principle. *Philosphy of Science*, 26:125–134, 1959.

[74] M. Bunge. Mach's critique of newtonian mechanics. *American Journal of Physics*, 34:585–596, 1966.

[75] M. Reinhardt. Mach's principle — A critical review. *Zeitschritte fur Naturforschung A*, 28:529–537, 1973.

[76] T. E. Phipps Jr. Should Mach's principle be taken seriously? *Speculations in Science and Technology*, 1:499–508, 1978.

[77] D. J. Raine. Mach's principle and space-time structure. *Reports on Progress in Physics*, 44:1151–1195, 1981.

[78] B. Friedlander and I. Friedlander. *Absolute oder Relative Bewegung?* Leonhard Simion, Berlin, 1896.

[79] B. Friedlaender and I. Friedlaender. Absolute or relative motion? In J. B. Barbour and H. Pfister, editors, *Mach's Principle, From Newton's Bucket to Quantum Gravity*, pages 114–119 and 309–311, Boston, 1995. Birkhäuser. Translated by J. B. Barbour.

[80] W. Hofmann. Motion and inertia. In J. B. Barbour and H. Pfister, editors, *Mach's Principle, From Newton's Bucket to Quantum Gravity*, pages 128–133, Boston, 1995. Birkhäuser. Translated by J. B. Barbour.

[81] H. Reissner. On the relativity of accelerations in mechanics. In J. B. Barbour and H. Pfister, editors, *Mach's Principle, From Newton's Bucket to Quantum Gravity*, pages 134–142, Boston, 1995. Birkhäuser. Translated by J. B. Barbour.

[82] H. Reissner. On a possibility of deriving gravitation as a direct consequence of the relativity of inertia. In J. B. Barbour and H. Pfister, editors, *Mach's Principle - From Newton's Bucket to Quantum Gravity*, pages 143–146, Boston, 1995. Birkhäuser. Translated by J. B. Barbour.

[83] E. Schrödinger. The possibility of fulfillment of the relativity requirement in classical mechanics. In J. B. Barbour and H. Pfister, editors, *Mach's Principle - From Newton's Bucket to Quantum Gravity*, pages 147–158, Boston, 1995. Birkhäuser. Translated by J. B. Barbour.

[84] E. Mach. *Popular Scientific Lectures*. Open Court, Chicago, 4th edition, 1910.

[85] A. Pais. *'Subtle is the Lord...' — The Science and the Life of Albert Einstein*. Oxford University Press, Oxford, 1982.

[86] A. Einstein. On the electrodynamics of moving bodies. In A. Einstein, H. A. Lorentz, H. Weyl and H. Minkowski, *The Principle of Relativity*, pages 35–65, New York, 1952. Dover.

[87] M. Faraday. *Experimental Researches in Electricity*, volume 45, pp. 257–866 of *Great Books of the Western World*. Encyclopaedia Britannica, Chicago, 1952.

[88] A. I. Miller. *Albert Einstein's Special Theory of Relativity*. Addison-Wesley, Reading, 1981.

[89] J. C. Maxwell. *A Treatise on Electricity and Magnetism*. Dover, New York, 1954.

[90] W. A. Rodrigues Jr. and E. C. de Oliveira. A comment on the twin paradox and the Hafele-Keating experiment. *Physics Letters A*, 140:479–484, 1989.

[91] W. A. Rodrigues Jr. and M. A. F. Rosa. The meaning of time in the theory of relativity and Einstein's later view on the twin paradox. *Foundations of Physics*, 19:705–724, 1989.

[92] T. E. Phipps, Jr. Clock rates in a machian universe. *Toth-Maatian Review*, 13:5910–5917, 1996.

[93] J. P. Wesley. *Selected Topics in Advanced Fundamental Physics*. Benjamin Wesley Publisher, Blumberg, 1991.

[94] S. A. Tolchelnikova-Murri. A new way to determine the velocity of the solar system. *Galilean Electrodynamics*, 3:72–75, 1992.

[95] H. C. Hayden. Special relativity: problems and alternatives. *Physics Essays*, 8:366–374, 1995.

[96] R. A. Monti. Theory of relativity: a critical analysis. *Physics Essays*, 9:238–260, 1996.

[97] I. B. Cohen. Roemer and the first determination of the velocity of light. *Isis*, 31:327–379, 1940.

[98] R. Taton (editor). *Roemer et la Vitesse de la Lumière*. J. Vrin, Paris, 1978.

[99] O. Roemer. The velocity of light. In W. F. Magie, editor, *A Source Book in Physics*, pages 335–337, New York, 1935. McGraw-Hill.

[100] J. Bradley. The discovery of the aberration of light. In H. Shapley and H. E. Howarth, editors, *A Source Book in Astronomy*, pages 103–108, New York, 1929. McGraw-Hill.

[101] J. Bradley. The velocity of light. In W. F. Magie, editor, *A Source Book in Physics*, pages 337–340, New York, 1935. McGraw-Hill.

[102] G. Sarton. Discovery of the aberration of light. *Isis*, 6:233–239, 1931.

[103] A. K. T. Assis and F. M. Peixoto. On the velocity in the Lorentz force law. *The Physics Teacher*, 30:480–483, 1992.

[104] E. T. Whittaker. *A History of the Theories of Aether and Electricity*, volume 1: *The Classical Theories*. Humanities Press, New York, 1973.

[105] J. J. Thomson. On the electric and magnetic effects produced by the motion of electrified bodies. *Philosophical Magazine*, 11:229–249, 1881.

[106] O. Heaviside. On the electromagnetic effects due to the motion of electrification through a dielectric. *Philosophical Magazine*, 27:324–339, 1889.

[107] H. A. Lorentz. *The Theory of Electrons*. Teubner, Leipzig, second edition, 1915. Reprinted in *Selected Works of H. A. Lorentz*, vol. 5, N. J. Nersessian (Ed.), (Palm Publications, Nieuwerkerk, 1987).

[108] A. O'Rahilly. *Electromagnetic Theory — A Critical Examination of Fundamentals*. Dover, New York, 1965.

[109] H. A. Lorentz. *Lectures on Theoretical Physics*, volume 3. MacMilan, London, 1931.

[110] H. A. Lorentz. Michelson's interference experiment. In A. Einstein, H. A. Lorentz, H. Weyl and H. Minkowski, *The Principle of Relativity*, pages 1–7, New York, 1952. Dover.

[111] S. A. Tolchelnikova-Murri. The Doppler observations of Venus contradict SRT. *Galilean Electrodynamics*, 4:3–6, 1993.

[112] W. A. Rodrigues Jr. and J. Y. Lu. On the existence of undistorted progressive waves (UPWs) of arbitrary speeds $0 \leq v < \infty$ in nature. *Foundations of Physics*, 27:435–508, 1997.

[113] A. Einstein. The foundations of the general theory of relativity. In A. Einstein, H. A. Lorentz, H. Weyl and H. Minkowski, *The Principle of Relativity*, pages 109–164, New York, 1952. Dover.

[114] A. Einstein. *The Meaning of Relativity*. Chapman and Hall, London, 1980.

[115] A. Einstein. Cosmological considerations on the general theory of relativity. In A. Einstein, H. A. Lorentz, H. Weyl and H. Minkowski, *The Principle of Relativity*, pages 175–188, New York, 1952. Dover.

[116] C. H. Brans. Mach's principle and the locally measured gravitational constant in general relativity. *Physical Review*, 125:388–396, 1962.

[117] H. Pfister. Dragging effects near rotating bodies and in cosmological models. In J. B. Barbour and H. Pfister, editors, *Mach's Principle - From Newton's Bucket to Quantum Gravity*, pages 315–331, Boston, 1995. Birkhäuser.

[118] H. Thirring. Über die wirkung rotierender ferner massen in der Einsteinschen gravitationstheorie. *Physikalische Zeitschrift*, 19:33–39, 1918.

[119] H. Thirring. Berichtigung zu meiner arbeit: 'Über die wirkung rotierender ferner massen in der Einsteinschen gravitationstheorie'. *Physikalische Zeitschrift*, 22:29–30, 1921.

[120] B. Mashhoon, F. H. Hehl, and D. S. Theiss. On the gravitational effects of rotating masses: the Thirring-Lense papers. *General Relativity and Gravitation*, 16:711–750, 1984.

[121] F. M. Peixoto and M. A. F. Rosa. On Thirring's approach to Mach's principle: criticisms and speculations on extensions of his original work. In P. S. Letelier and W. A. Rodrigues Jr., editors, *Gravitation: The Spacetime Structure*, pages 172–178, Singapore, 1994. World Scientific. Proceedings of the 8th Latin American symposium on relativity and gravitation.

[122] L. Bass and F. A. E. Pirani. On the gravitational effects of distant rotating masses. *Philosophical Magazine*, 46:850–856, 1955.

[123] D. R. Brill and J. M. Cohen. Rotating masses and their effect on inertial frames. *Physical Review*, 143:1011–1015, 1966.

[124] J. M. Cohen and D. R. Brill. Further examples of "machian" effects of rotating bodies in general relativity. *Nuovo Cimento B*, 56:209–219, 1968.

[125] E. Schrödinger. Die erfüllbarkeit der relativitätsforderung in der klassischen mechanik. *Annalen der Physik*, 77:325–336, 1925.

[126] H.-H. von Borzeszkowski and H.-J. Treder. Mach-Einstein doctrine and general relativity. *Foundations of Physics*, 26:929–942, 1996.

[127] Y. Bozhkov and W. A. Rodrigues Jr. Mass and energy in general relativity. *General Relativity and Gravitation*, 27:813–819, 1995.

[128] E. Mach. *The Principles of Physical Optics — An Historical and Philosophical Treatment.* E. P. Dutton and Company, New York, 1926. Reprinted by Dover, New York, in 1953.

[129] J. Blackmore. Ernst Mach leaves 'the church of physics'. *British Journal for the Philosophy of Science*, 40:519–540, 1989.

[130] A. K. T. Assis. Deriving gravitation from electromagnetism. *Canadian Journal of Physics*, 70:330–340, 1992.

[131] A. K. T. Assis. Gravitation as a fourth order electromagnetic effect. In T. W. Barrett and D. M. Grimes, editors, *Advanced Electromagnetism: Foundations, Theory and Applications*, pages 314–331, Singapore, 1995. World Scientific.

[132] A. K. T. Assis. On Mach's principle. *Foundations of Physics Letters*, 2:301–318, 1989.

[133] P. A. M. Dirac. A new basis for cosmology. *Proceedings of the Royal Society of London A*, 165:199–208, 1938.

[134] G. Borner. *The Early Universe – Facts and Fiction*. Springer, Berlin, 1988.

[135] D. Roscoe. Gravity out of inertia. *Apeiron*, 9-10:54–61, 1991.

[136] A. K. T. Assis. On Hubble's law of redshift, Olbers' paradox and the cosmic background radiation. *Apeiron*, 12:10–16, 1992.

[137] M. C. D. Neves and A. K. T. Assis. The Compton effect as an explanation for the cosmological redshift. *Quarterly Journal of the Royal Astronomical Society*, 36:279–280, 1995.

[138] A. K. T. Assis and M. C. D. Neves. The redshift revisited. *Astrophysics and Space Science*, 227:13–24, 1995. This paper was also published in *Plasma Astrophysics and Cosmology*, A. L. Peratt (ed.), (Kluwer Academic Publishers, Dordrecht, 1995), pp. 13-24.

[139] E. Regener. Der energiestrom der ultrastrahlung. *Zeitschrift fur Physik*, 80:666–669, 1933.

[140] E. Regener. The energy flux of cosmic rays. *Apeiron*, 2:85–86, 1995.

[141] W. Nernst. Weitere prüfung der annahne eines stationären zustandes im weltall. *Zeitschrift fur Physik*, 106:633–661, 1937.

[142] W. Nernst. Die strahlungstemperatur des universums. *Annalen der Physik*, 32:44–48, 1938.

[143] W. Nernst. Further investigation of the stationary universe hypothesis. *Apeiron*, 2:58–71, 1995.

[144] W. Nernst. The radiation temperature of the universe. *Apeiron*, 2:86–87, 1995.

[145] E. Finlay-Freundlich. Über die rotverschiebung der spektrallinien. In *Nachrichten der Akademie der Wissenschaften in Göttingen Mathematisch-Physikalische Klasse, N. 7*, pages 95–102, Göttingen, 1953. Vandenhoeck & Ruprecht.

[146] E. Finlay-Freundlich. Red-shifts in the spectra of celestial bodies. *Proceedings of the Physical Society A*, 67:192–193, 1954.

[147] E. Finlay-Freundlich. Red shifts in the spectra of celestial bodies. *Philosophical Magazine*, 45:303–319, 1954.

[148] M. Born. Theoretische bemerkungen zu Freundlich's formel für die stellare rotverschiebung. In *Nachrichten der Akademie der Wissenschaften in Göttingen Mathematisch-Physikalische Klasse, N. 7*, pages 102–108, Göttingen, 1953. Vandenhoeck & Ruprecht.

[149] M. Born. On the interpretation of Freundlich's red-shift formula. *Proceedings of the Physical Society A*, 67:193–194, 1954.

[150] L. de Broglie. Sur le déplacement des raies émises par un objet astronomique lointain. *Comptes Rendues de l'Academie des Sciences de Paris*, 263:589–592, 1966.

[151] H. C. Arp, C. R. Keys, and K. Rudnicki. *Progress in New Cosmologies - Beyond the Big Bang*. Plenum Press, New York, 1993.

[152] G. Reber. Endless, boundless, stable universe. *University of Tasmania occasional paper*, 9:1–18, 1977.

[153] G. Reber. Intergalactic plasma. *IEEE Transactions on Plasma Science*, PS-14:678–682, 1986.

[154] P. Marmet and G. Reber. Cosmic matter and the nonexpanding universe. *IEEE Transactions on Plasma Science*, 17:264–269, 1989.

[155] A. Ghosh. Velocity dependent inertial induction: An extension of Mach's principle. *Pramana Journal of Physics*, 23:L671–L674, 1984.

[156] A. Ghosh. Velocity-dependent inertial induction and secular retardation of the earth's rotation. *Pramana Journal of Physics*, 26:1–8, 1986.

[157] A. Ghosh. Velocity dependent inertial induction - a case for experimental observation. *Apeiron*, 3:18–23, 1988.

[158] A. Ghosh. Astrophysical and cosmological consequences of velocity-dependent inertial induction. In H. Arp, R. Keys, and K. Rudnicki, editors, *Progress in New Cosmologies: Beyond the Big Bang*, pages 305–326, New York, 1993. Plenum Press.

[159] J.-C. Pecker. How to describe physical reality? *Apeiron*, 2:1–12, 1988.

[160] J.-C. Pecker and J.-P. Vigier. A possible tired-light mechanism. *Apeiron*, 2:13–15, 1988.

[161] P. Marmet. A new mechanism to explain observations incompatible with the big bang. *Apeiron*, 9-10:45–53, 1991.

[162] P. A. LaViolette. Is the universe really expanding? *Astrophysical Journal*, 301:544–553, 1986.

[163] P. Gen. New insight into Olbers' and Seeliger's paradoxes and the cosmic background radiation. *Chinese Astronomy and Astrophysics*, 12:191–196, 1988.

[164] T. Jaakkola. Mach's principle and properties of local structure. *Apeiron*, 1:5–12, 1987.

[165] T. Jaakkola. The cosmological principle: theoretical and empirical foundations. *Apeiron*, 4:8–31, 1989.

[166] T. Jaakkola. On reviving tired light. *Apeiron*, 6:5–6, 1990.

[167] T. Jaakkola. Electro-gravitational coupling: Empirical and theoretical arguments. *Apeiron*, 9-10:76–90, 1991.

[168] K. Rudnicki. The importance of cosmological principles for research in cosmology. *Apeiron*, 4:1–7, 1989.

[169] K. Rudnicki. What are the empirical bases of the Hubble law? *Apeiron*, 9-10:4–7, 1991.

[170] E. Hubble. *The Observational Approach to Cosmology*. Clarendon Press, Oxford, 1937.

[171] E. Hubble. The problem of the expanding universe. *American Scientist*, 30:99–115, 1942.

[172] E. Hubble. *The Realm of the Nebulae*. Dover, New York, 1958.

[173] A. K. T. Assis and M. C. D. Neves. History of the 2.7 K temperature prior to Penzias and Wilson. *Apeiron*, 2:79–84, 1995.

[174] R. A. Alpher and R. Herman. Evolution of the universe. *Nature*, 162:774–775, 1948.

[175] R. A. Alpher and R. Herman. Remarks on the evolution of the expanding universe. *Physical Review*, 75:1089–1095, 1949.

[176] G. Gamow. *The Creation of the Universe*. Viking Press, New York, revised edition, 1961.

[177] C. E. Guillaume. La température de l'espace. *La Nature*, 24, series 2:210–211, 234, 1896.

[178] A. S. Eddington. *The Internal Constitution of the Stars*. Cambridge University Press, Cambridge, 1988. Reprint of 1926 edition.

[179] G. Herzberg. *Molecular Spectra and Molecular Structure*, volume 1: Spectra of Diatomic Molecules. D. Van Nostrand, New York, 1941.

[180] R. A. Clemente and A. K. T. Assis. Two-body problem for Weber-like interactions. *International Journal of Theoretical Physics*, 30:537–545, 1991.

[181] A. K. T. Assis. Circuit theory in Weber electrodynamics. *European Journal of Physics*, 18:241–246, 1997.

[182] P. Eby. Machian mechanics with isotropic inertia. *General Relativity and Gravitation*, 29:621–635, 1997.

[183] Maurice Allais. *L'Anisotropie de L'Espace - Les Données de L'Expérience*. Clément Juglar, Paris, 1997.

[184] J. P. Wesley. Weber electrodynamics, Part III. Mechanics, gravitation. *Foundations of Physics Letters*, 3:581–605, 1990.

[185] S. Ragusa. Gravitation with a modified Weber force. *Foundations of Physics Letters*, 5:585–589, 1992.

[186] F. Bunchaft and S. Carneiro. Weber-like interactions and energy conservation. *Foundations of Physics Letters*, 10:393–401, 1997.

[187] G. Cocconi and E. Salpeter. A search for anisotropy of inertia. *Nuovo Cimento*, 10:646–651, 1958.

[188] G. Cocconi and E. Salpeter. Upper limit for the anisotropy of inertia from the Mössbauer effect. *Physical Review Letters*, 4:176–177, 1960.

[189] C. W. Sherwin *et al.* Search for the anisotropy of inertia using the Mössbauer effect in Fe^{57}. *Physical Review Letters*, 4:399–401, 1960.

[190] V. W. Hughes *et al.* Upper limit for the anisotropy of inertial mass from nuclear resonance experiments. *Physical Review Letters*, 4:342–344, 1960.

[191] R. W. P. Drever. A search for anisotropy of inertial mass using a free precession technique. *Philosophical Magazine*, 6:683–687, 1961.

[192] R. H. Dicke. Experimental tests of Mach's principle. *Physical Review Letters*, 7:359, 1961. Reprinted in: R. H. Dicke, *The Theoretical Significance of Experimental Relativity*, (Gordon and Breach, New York, 1964), pp. 31-34.

[193] W. F. Edwards. Inertia and an alternative approach to the theory of inter-actions. *Proceedings of the Utah Academy of Science, Arts, and Letters*, 51, Part 2:1–7, 1974.

[194] P. Eby. Gyro precession and Mach's principle. *General Relativity and Gravitation*, 11:111–117, 1979.

[195] I. B. Cohen. *The Newtonian Revolution*. Cambridge University Press, Cambridge, 1980.

[196] I. B. Cohen. Newton's discovery of gravity. *Scientific American*, 244:122–133, 1981.

[197] A. Coulomb. First memoir on electricity and magnetism. In W. F. Magie, editor, *A Source Book in Physics*, pages 408–413, New York, 1935. McGraw-Hill.

[198] A. Coulomb. Second memoir on electricity and magnetism. In W. F. Magie, editor, *A Source Book in Physics*, pages 413–420, New York, 1935. McGraw-Hill.

[199] A. M. Ampère. *Théorie Mathématique des Phénomenes Électrodynami-ques Uniquement Déduite de l'Expérience*. Blanchard, Paris, 1958.

[200] R. A. R. Tricker. *Early Electrodynamics — The First Law of Circulation*. Pergamon, New York, 1965.

[201] W. Weber. Elektrodynamische maassbestimmungen über ein allgemeines grundgesetz der elektrischen wirkung. *Abhandlungen bei Begründung der Königl. Sächs. Gesellschaft der Wissenschaften am Tage der zweihundert-jährigen Geburtstagfeier Leibnizen's herausgegeben von der Fürstl. Jablo-nowskischen Gesellschaft (Leipzig)*, pages 211–378, 1846. Reprinted in Wilhelm Weber's *Werke* (Springer, Berlin, 1893), vol. 3, pp. 25–214.

[202] W. Weber. Elektrodynamische maassbestimmungen. *Annalen der Physik*, 73:193–240, 1848.

[203] W. Weber. On the measurement of electro-dynamic forces. In R. Tay-lor, editor, *Scientific Memoirs*, vol. 5, pages 489–529, New York, 1966. Johnson Reprint Corporation.

[204] A. E. Woodruff. The contributions of Hermann von Helmholtz to elec-trodynamics. *Isis*, 59:300–311, 1968.

[205] M. N. Wise. German concepts of force, energy, and the electromagnetic ether: 1845–1880. In G. N. Canter and M. J. S. Hodge, editors, *Conceptions of Ether — Studies in the History of Ether Theories 1740–1900*, pages 269–307, Cambridge, 1981. Cambridge University Press.

[206] T. Archibald. Energy and the mathematization of electrodynamics in Germany, 1845-1875. *Archives Internationales d'Histoire des Sciences*, 39:276–308, 1989.

[207] P. Graneau. *Ampere-Neumann Electrodynamics of Metals*. Hadronic Press, Nonantum, 1985.

[208] Marcelo A. Bueno and A. K. T. Assis. A new method for inductance calculations. *Journal of Physics D*, 28:1802–1806, 1995.

[209] C. G. Darwin. The dynamical motions of charged particles. *Philosophical Magazine*, 39:537–551, 1920.

[210] J. D. Jackson. *Classical Electrodynamics*. John Wiley, New York, second edition, 1975.

[211] W. Weber and R. Kohlrausch. Ueber die elektricitätsmenge, welche bei galvanischen strömen durch den querschnitt der kette fliesst. *Annalen der Physik*, 99:10–25, 1856.

[212] F. Kirchner. Determination of the velocity of light from electromagnetic measurements according to W. Weber and R. Kohlrausch. *American Journal of Physics*, 25:623–629, 1957.

[213] L. Rosenfeld. The velocity of light and the evolution of electrodynamics. *Il Nuovo Cimento*, Supplement to vol. 4:1630–1669, 1957.

[214] L. Rosenfeld. Kirchhoff, Gustav Robert. In C. C. Gillispie, editor, *Dictionary of Scientific Biography*, vol. 7, pages 379–383, New York, 1973. Scribner.

[215] A. E. Woodruff. Weber, Wilhelm Eduard. In C. C. Gillispie, editor, *Dictionary of Scientific Biography*, vol. 14, pages 203–209, New York, 1976. Scribner.

[216] P. M. Harman. *Energy, Force, and Matter – The Conceptual Development of Nineteenth-Century Physics*. Cambridge University Press, Cambridge, 1982.

[217] C. Jungnickel and R. McCormmach. *Intellectual Mastery of Nature — Theoretical Physics from Ohm to Einstein*, volume 1–2. University of Chicago Press, Chicago, 1986.

[218] L. Hecht. The significance of the 1845 Gauss-Weber correspondence. *21st Century*, 9(3):22–34, 1996.

[219] H. von Helmholtz. On the theory of electrodynamics. *Philosophical Magazine*, 44:530–537, 1872.

[220] P. Gerber. Die räumliche und zeitliche ausbreitung der gravitation. *Zeitschrift fur Mathematik und Physik II*, 43:93–104, 1898.

[221] P. Gerber. Die fortpflanzungsgeschwindigkeit der gravitation. *Annalen der Physik*, 52:415–444, 1917.

[222] P. B. Eby. On the perihelion precession as a Machian effect. *Lettere al Nuovo Cimento*, 18:93–96, 1977.

[223] H. Poincaré. Les limites de la loi de Newton. *Bulletin Astronomique*, 17:121–269, 1953.

[224] H. Seeliger. Bemerkung zu P. Gerbers aufsatz: 'Die fortplanzungsgeschwindligkeit der gravitation'. *Annalen der Physik*, 53:31–32, 1917.

[225] H. Reissner. Über die relativität der beschleunigungen in der mechanik. *Physikalishe Zeitschrift*, 15:371–375, 1914.

[226] H. Reissner. Über eine möglichkeit die gravitation als unmittelbare folge der relativität der trägheit abzuleiten. *Physikalishe Zeitschrift*, 16:179–185, 1915.

[227] H. Reissner. Über die Eigengravitation des elektrischen Fedels nach der Einsteinschen Theorie. *Annalen der Physik*, 50:106–120, 1916.

[228] E. Schrödinger. *Collected Papers*, volume 2. Austrian Academy of Sciences, Vienna, 1984.

[229] O. Hittmair. Schrödinger's unified theory seen 40 years later. In C. W. Kilmister, editor, *Schrödinger - Centenary Celebration of a Polymath*, pages 165–175. Cambridge University Press, Cambridge, 1987.

[230] E. Schrödinger. *Expanding Universes*. Cambridge University Press, Cambridge, 1957.

[231] J. Mehra. Erwin Schrödinger and the rise of wave mechanics. II. The creation of wave mechanics. *Foundations of Physics*, 17:1141–1188, 1987.

[232] A. L. Xavier Jr. and A. K. T. Assis. O cumprimento do postulado de relatividade na mecânica clássica - uma tradução comentada de um texto de Erwin Schrödinger sobre o princípio de Mach. *Revista da Sociedade Brasileira de História da Ciência*, 12:3–18, 1994.

[233] G. B. Brown. *Retarded Action-at-a-Distance*. Cortney Publications, Luton, 1982.

[234] J. B. Barbour. Relative-distance Machian theories. *Nature*, 249:328–329, 1974. Misprints corrected in Nature, vol. 250, p. 606 (1974).

[235] J. B. Barbour and B. Bertotti. Gravity and inertia in a Machian framework. *Nuovo Cimento B*, 38:1–27, 1977.

[236] J. B. Barbour and B. Bertotti. Mach's principle and the structure of dynamical theories. *Proceedings of the Physical Society of London A*, 382:295–306, 1982.

[237] Marcelo Bueno and A. K. T. Assis. *Cálculo de Indutância e de Força em Circuitos Elétricos*. Editora da UFSC/Editora da UEM, Florianópolis/Maringá, 1998. ISBN: 85-328-0119-6.

[238] P. Graneau. The riddle of inertia. *Electronics and Wireless World*, 96:60–62, 1990.

[239] P. Graneau. Far-action versus contact action. *Speculations in Science and Technology*, 13:191–201, 1990.

[240] P. Graneau. Interconnecting action-at-a-distance. *Physics Essays*, 3:340–343, 1990.

[241] P. Graneau. Some cosmological consequences of Mach's principle. *Hadronic Journal Supplement*, 5:335–349, 1990.

[242] P. Graneau. Has the mystery of inertia been solved? In U. Bartocci and J. P. Wesley, editors, *Proceedings of the Conference on Foundations of Mathematics and Physics*, pages 129–136, Blumberg, Germany, 1990. Benjamin Wesley Publisher.

[243] P. Graneau and N. Graneau. *Newton Versus Einstein – How Matter Interacts with Matter*. Carlton Press, New York, 1993.

[244] A. Zylbersztajn. Newton's absolute space, Mach's principle and the possible reality of fictitious forces. *European Journal of Physics*, 15:1–8, 1994.

Index